Lasers and Electro-Optics Research and Technology

TWO-DIMENSIONAL PERIODIC NANOSCALE PATTERNING OF SOLID SURFACES BY FOUR-BEAM STANDING WAVE EXCIMER LASER LITHOGRAPHY

LASERS AND ELECTRO-OPTICS RESEARCH AND TECHNOLOGY

Performance Monitoring in Advanced Optical Fiber Networks
Lianshan Yan (Author)
2009. 978-1-60692-645-1

Fiber Lasers: Research, Technology and Applications
Masato Kimura (Editor)
2009. 978-1-60692-896-7

High-Power and Femtosecond Lasers: Properties, Materials and Applications
Paul-Henri Barret and Michael Palmer (Editors)
2009. 978-1-60741-009-6

From Femto-to Attoscience and Beyond
Janina Marciak-Kozlowska and Miroslaw Kozlowski (Authors)
2009. 978-1-60741-164-2

Optical Solitons in Nonlinear Micro Ring Resonators: Unexpected Results and Applications
Nithiroth Pornsuwancharoen, Jalil Ali and Preecha Yupapin (Authors)
2009. 978-1-60741-342-4

Phase Transitions Induced by Short Laser Pulses
Georgy A. Shafeev (Editor)
2009. 978-1-60741-590-9

Laser Beams: Theory, Properties and Applications
Maxim Thys and Eugene Desmet (Editors)
2010. 978-1-60876-266-8

Quantum Well Structures for Infrared Photodetection
Wei Shi and D.H. Zhang (Authors)
2010. 978-1-61668-368-9

Optical and Electro-Optical Properties of Liquid Crystals: Nematic and Smecic Phases
Minko Parvanov Petrov (Author)
2010. 978-1-61668-360-3

Two-Dimensional Periodic Nanoscale Patterning of Solid Surfaces by Four-Beam Standing Wave Excimer Laser Lithography
D. O. Filatov (Author)
2010. 978-1-61668-674-1

New Organic Semiconductors for Applications in Organic Electronics
Chunyan Du and Yunqi Liu (Authors)
2010. 978-1-61668-764-9

Lasers and Electro-Optics Research and Technology

TWO-DIMENSIONAL PERIODIC NANOSCALE PATTERNING OF SOLID SURFACES BY FOUR-BEAM STANDING WAVE EXCIMER LASER LITHOGRAPHY

D. O. FILATOV

Nova Science Publishers, Inc.
New York

For permission to use material from this book please contact us:
Telephone 631-231-7269; Fax 631-231-8175
Web Site: http://www.novapublishers.com

Library of Congress Cataloging-in-Publication Data

Available upon Request

ISBN: 978-1-61668-674-1

Published by Nova Science Publishers, Inc. † New York

CONTENTS

PREFACE

This book is devoted to a promising technology for fabrication of two-dimensional, nanoperiodic structures on solid surfaces by four-beam standing wave pulsed laser modification. In this method, a large number of spots (up to 10^7) less than 100 nm in size heated up to several thousand degrees are formed on the modified surface as a result of the interference of the four laser beams incident on the object coherently. Besides a direct laser ablation, various nonlinear mechanisms of surface modification may take place in the standing wave maxima. This way, a nearly perfect two-dimensional pattern can be formed on the modified surface; the pitches of the pattern in both directions can be adjusted easily within the range from a few microns to tens of nanometers by simply varying the beams' incidence angles. This method being developed extensively at present is very promising for application in microelectronics, data storage, integrated optics (including fabrication of the Bragg mirrors, antireflection coatings, two-dimensional photonic crystals, etc.), and is considered to be a promising alternative to X-ray lithography in a wide range of applications. The most important advantage of the method is that the nanoperiodic pattern can be formed in a single step (actually, at single laser pulse) by direct laser ablation of the material. However, the method could be applied as well to patterning a resist deposited onto the processed substrate to serve as a lithographic mask. In this book, we present some experimental results on patterning various materials, such as metal films on semiconductor and glass substrates, polyimide, titanium dioxide, and even such a hard material as diamond by direct laser ablation using a XeCl excimer laser. Patterning a resist film, the phase masks on fused silica and the antireflection structures on Si and silica glass have been fabricated. Also, using the effect of stimulation of self-assembling of the semiconductor nanostructures by standing wave laser irradiation discovered by the authors, we have fabricated the ordered arrays of nanoislands on the surface of single crystal Si and GaAs and of GaAs/$In_xGa_{1-x}As$/GaAs(001) heterostructures with the diameter ranging from 5 to 60 nm.

The authors gratefully acknowledge the financial support of the studies presented in this book by European Community FP-6 IST-4 No.027976 “Development of Lithography Technology for Nanoscale Structuring of Materials Using Laser Beam Interference (DELILA)”, by the Russian Academy of Science (Division of Physics Sciences) under Nonlinear Optics of Unique Laser Systems Program, by the Russian Foundation for Basic Research (L-EN 96-15-96594, 98-02-16306, and 02-02-17139), and by the Joint Russian – American Program “Basic Research and Higher Education”(BRHE) funded in parity by U.S. Civilian Research and Development Foundation (CRDF) and by the Ministry of Education, Russian Federation (REC-NN-001). Also the authors are grateful to Dmitry Vorontsov, Dept. of Physics, N. I. Lobachevskii University of Nizhny Novgorod for the assistance in the preparation of the manuscript for publication.

Chapter 1

INTRODUCTION

The periodic submicron structures have found a wide range of applications in science and technology [1] Among the applications of these structures, one could mention Bragg mirrors, photonic crystals, data storage, etc. For formation of the periodic arrays of the nanometer-sized elements, E-Beam Lithography and X-Ray Lithography are used the most often. One of the promising methods for forming the submicron periodic arrays on the solid surfaces is Standing Wave or Interference Lithography [2]. In this method, the periodic pattern on the surface to be processed is provided by the interference of two (or more) coherent laser beams incident onto the surface. In the case of two beams, the interference pattern is suitable for forming various gratings, first of all, for the integrated and fiber optics [3,4], including the Bragg mirrors, beam couplers, multiplexers, etc. As follows from a simple geometrical consideration presented in figure 1, the period of the standing wave in the symmetric scheme $d = \lambda/2\sin\vartheta$ where λ is the wavelength of the radiation used and ϑ is the angle of incidence. So far, one can vary the period of the grating formed within a rather wide range simply adjusting ϑ. The minimum $d_{min} = \lambda/2$ is reached when $\vartheta \rightarrow \pi/2$. Using an immersion reduces d_{min} n times, where n is the refractive index of the immersion medium.

In the four-beam Standing Wave Lithography, the material to be patterned is subjected to the two pairs of the coherent laser beams incident on the surface under relatively large angles. So far, the energy of the laser beams is localized in a huge number of spots of $\sim \lambda/4$ in size. When the excimer ultraviolet (UV) lasers are used, the light can be localized in the spots of less than 40 nm in size. The use of the 4th (or even the 5th) harmonic of a solid state laser radiation represents an alternative approach.

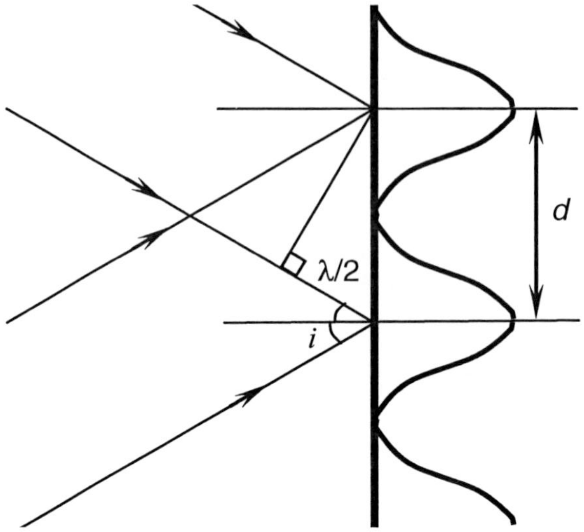

Figure 1. A schematic representation of two-beam interference in the symmetric scheme.

It should be noted that E-Beam Lithography and X-Ray Lithography exceed Standing Wave Lithography in the spatial resolution. The forthcoming generation of the semiconductor devices will be based on sub-50-nm critical dimensions [5]. These critical dimensions are planned to be achieved using the nonlinear effects in Extreme UV (EUV) Lithography. The above listed techniques have higher spatial resolution limit than UV Standing Wave Lithography. In addition, E-Beam Lithography is featured not only by the lowest critical spatial dimensions reached to date, but also by almost complete pattern flexibility: it allows writing any nanoscale structures within the limits imposed by proximity effects. Therefore, E-Beam Lithography is the most suitable method for nanoscience research purposes. X-Ray and EUV Lithographies succeed the flexibility from E-Beam Lithography in a considerable amount, since the photomasks are produced by the latter method mostly. In this scope, Standing Wave Lithography offers a very poor choice of the patterns (compared to the E-Beam Lithography or to X-Ray Lithography), which it is capable to produce. This choice is limited to the rather simple periodic arrays of stripes (in the case of Two Beam Standing Wave Lithography) or to the square (or the rhombic) periodic arrays of the round or oval elements (in the case of Four Beam Standing Wave Lithography). Nonetheless, there are many fields of application, which require strictly periodic one- or two-dimensional arrays of the similar elements of relatively large areas (of several square centimeters). Standing Wave Lithography can realize its advantages in these areas in full amount. These are, first of all, the perfect periodicity of the pattern produced and higher productivity compared to E-Beam Lithography. The latter is not a viable method for addressing such tasks because of its poor productivity that is a result of its serial point-by-point writing principle. So far, it takes too long time to record a relatively large-area nanoscale structure or mask.

In this book we present a number of experimental results on patterning various materials by Two- and Four Beam Standing Wave Laser Lithography obtained using a powerful UV XeCl excimer pulsed laser with the emission wavelength $\lambda = 308$ nm. The apparatus used in these experiments is described in Chapter II. Also, some features of the interference of four laser beams and the mechanisms of the surface modification under pulsed interfering UV excimer laser radiation are discussed briefly. Finally, the heat dissipation in a thin film on a semi-infinite heat conducting substrate subjected to a spatially periodic surface heat source (i.e. the standing wave of pulsed laser radiation) [6] is considered in details in Chapter II. It was shown that the process of the heat release governs all the processes of the film modification (first of all, the direct ablation of the film material), and is important for interpreting the experimental results presented in the following chapters.

There are two main techniques of Standing Wave Laser Lithography. In the first technique, the surface to be patterned is covered by a photoresist and exposed to the interfering laser radiation. Then the exposed resist undergoes development followed by further processing as in the routine UV lithography [7,8,9]. The possibility of easy incorporation into a standard UV photolithographic process is a considerable advantage of Standing Wave Lithography. An alternative technique is a direct (a single shot) modification of the patterned surface by laser irradiation [10,11,12,13]. In the studies presented in this book, both techniques have been applied. In the first technique, both continuous wave (CW) and pulsed lasers can be employed to expose the resist, whereas the second one needs high-power pulsed lasers. The high sensitivity of the photo-resists makes it possible to use relatively low pulse energy densities and therefore to spread the processed area up to more than 10 cm in diameter [2,7,9]. This approach is promising for forming the large area regular arrays of the nanostructures (e.g., for flat panel displays [14]). However, this multistage process is rather complicated and expensive.

In the experiments presented in this book, we have applied this technique to pattern the surface of fused silica [15]. Also, we have used two-beam Standing Wave Laser Lithography to form the one-dimensional phase masks on fused silica and on Si and the antireflection gratings on Si [16]. These objects are promising for applications in planar integrated optics [17,18]. The reason for using the photoresist was as follows. The integrated optics applications listed above require high enough aspect ratio of the surface relief that could hardly be obtained by direct laser ablation. The results of these studies are presented in Chapter III.

The second approach enables one to form sub-100-nm periodic arrays of the surface nanostructures in a single step (actually, at a single shot) in a variety of environments, including vacuum and ambient air. In Chapter IV, we present the results of the direct four-beam pulsed laser modification of the Feodificat on the Si substrates [6], of the polyimide and titanium dioxide films on the fused silica substrates, and of even such a hard material as diamond [15]. The submicron periodic regular arrays of the nanodimensional features of the surface morphology were obtained. Such structures are interesting in both fundamental and applied aspects. The Fe—C films are promising for the magnetic data storage applications. Also, investigation of the effect of the pulse energy density and duration on the resulting patterned film morphology turned to be useful to verify the theory of the heat release in the thin films on various substrates presented in Chapter II. The polyimide films are promising as the high contrast photoresists useful for application in microelectronic technology and in integrated optics for various purposes. It was demonstrated that the mechanisms of the polyimide film surface modification by the standing wave pulsed laser radiation are different for various values of the beam energy density. The titanium dioxide films are promising for the development of the efficient and cost-effective solar cells, for the bio catalysis, etc. Diamond has been considered to be a promising wide gap semiconductor material for several decades. One of its features is an extreme radiation hardness that makes it especially promising for application in nuclear power engineering, in high energy physics, in the space applications, etc. At present, the epitaxial films of the artificially synthesized diamond of nanometer scale thickness are considered to be a promising material for future nanoelectronic integrated circuits (ICs), first of all, due to high thermal conductivity of diamond. On the other hand, the hardness of diamond caused considerable problems in processing these films. The results presented in this book demonstrate the possibility of direct patterning of diamond by Standing Wave Laser Ablation. Besides, this technique can be applied to secret marking of the diamonds used in jewelry.

Investigations and applications of the self-assembling of nanostructures is now one of the dominant trends in the development of nanotechnology [19].

Starting from self-assembling of InAs/GaAs(001) quantum dots (QDs) [20] and GeSi/Si(001) nanoislands [21] driven by the elastic strain relaxation in the process of the heteroepitaxy of the highly mismatched materials [22] in the early 1990s, to date it has grown into a broad field of science and technology promising a wide range of applications in nanoelectronics [23], nanophotonics [24], etc. In recent years, extensive investigations have been devoted to finding the conditions for self-assembling of the spatially-ordered arrays of the self-assembled nanoislands in various material systems [25,26,27]. As a result of these studies, some self-assembling mechanism stimulating improvement in the size uniformity of the islands and their arrangement in regular arrays on the growth surface have been discovered [28,29]. Nevertheless, the problem of forming the uniform regular arrays of self-assembled semiconductor nanoisland still remains a vital one. Among various approaches to the solution of this problem, one could mention substrate pre-patterning [30] using lithography [31,32], masking [33], growth on the vicinal surfaces [34], using the artificial substrates such as Ge_xSi_{1-x}/Si(001) buffers for growing spatially-ordered GeSi nanoislands [35], etc. Pedraza et al. have studied the self-assembling of the nanoislands on the surface of Si substrate under high-power UV pulsed laser radiation in the regime of the surface wave excitation [36]. The self-assembling effect has been observed after more than 200 laser pulses with the pulse energy density ~ 1 J·cm^{-2}.

Recently a novel approach to forming near perfectly ordered periodic arrays of self-assembled semiconductor nanoislands has been proposed by the authors of the present book [37]. This is the self assembling of the semiconductor nanoislands under four-beam standing wave-pulsed laser radiation. In Chapter V the experimental results on the formation of the two-dimensional regular periodic arrays of the semiconductor nanoislands on the single crystal Si(001) and GaAs(001) substrates through self-assembling stimulated by the standing wave pulsed XeCl excimer laser radiation are presented. The surface stresses due to the local thermal expansion of the single crystal surface in the standing wave maxima were considered to be the driving force for the ordered nanoisland array formation. In addition, the results on the four beam standing wave pulsed laser modification of a GaAs/InGaAs/GaAs(001) strained heterostructures grown by Molecular Beam Epitaxy (MBE) are presented in Chapter V. The complex two-dimensional periodic nanoscale patterns were obtained. Within each cell of the standing wave induced pattern, a number of self-assembled nanoislands having a bimodal size distribution had been observed. The islands of smaller size (≈ 5 nm in diameter) were concentrated in the center of each cell whereas the larger islands (≈ 15 nm in diameter) were located at the boundaries of the cells. These results were explained by the interaction of the intrinsic elastic strain in the GaAs/InGaAs/GaAs heterostructure with the standing wave-induced thermal field induced strain.

Chapter 2

STANDING WAVE PULSED LASER MODIFICATION OF THE SOLID SURFACES: APPARATUS AND BASICS

A. APPARATUS

The techniques for laser modification of the solid surfaces can be classified into the following three groups according to the number of the laser beams employed in the processing. The methods, where a single focused laser beam is employed in, are the most developed ones [38]. The nonlinear mechanisms of modification of the solid surfaces under the focused laser beams provide the possibility to reduce the modified area down to the sizes comparable to the wavelength of the laser radiation used [39]. Using laser beam scanning and/or the motorized stages one can produce any complex structure including the Bragg mirrors for the infrared (IR) and visible wavelength bands. The main drawback of this approach is poor productivity that is an attribute of all technologies based on the scanning principle.

Another approach to forming the one-dimensional Bragg mirrors, gratings, etc. is using the interference of two coherent laser beams [40,41,42]. This way, an almost perfect periodic pattern over the area of several centimeters in diameter can be formed.

Two main experimental schemes for Standing Wave Lithography are presented in figure 2 [2]. The simplest setup [figure 2(a)] is a simple corner cube scheme (a Fresnel mirror) where the right and left halves of the input laser beam (already collimated and expanded) are folded onto each other. In this geometry, the incidence angle $\vartheta = 45°$, so far the dimensions of the beams incident onto the processed wafer and onto the mirror are equal to each other. The main advantages of this setup are the simplicity and the rigidness of the optical scheme as a whole. Expanding the input laser beam allows quite large processed areas. In this case, the latter is limited just by the size of the folding mirror. However, this setup requires the input laser beams with a rather high transverse coherence. The larger the processed area, the more strict requirements onto the transverse coherence of the input laser beam imposed. The laser beams provided by the single-mode TEM_{00} Ar^{+} lasers had given the satisfactory results [2].

On the other hand, the excimer lasers are featured (as a rule) by a large number of the transverse modes and, consequently, by a poor transverse coherence. So far, the Fresnel's mirror setup appears to be not suitable for the excimer laser sources, since the two halves of the input laser beams are not, in general, coherent mutually. An alternative experimental configuration, which solves the problem of the transverse coherence (so-called interferometric scheme), is presented in figure 2(b). In this case, the input beam is split in half and consequently folded onto itself. As it follows from the schematic representation in figure 2(b), in this scheme, any offset within the transverse coherence length remains constant. On the other hand, there appears the requirement of an essential longitudinal coherence (or, in other words, on a narrow enough spectral width) of the laser emission since the distances from the beam splitter to the processed wafer plane vary for the two beams across the illuminated field.

The state-of-art lithography-optimized excimer lasers provide very narrow line widths less than 1 pm (that corresponds to the longitudinal coherence length $\lambda^2/\Delta\lambda > 4$ cm, making large-area (at least several square centimeters) Standing Wave Lithography straightforward.

The interference of three and more coherent laser beams incident onto the surface to be modified in difference planes opens a way to create large two dimensional arrays of the submicron elements [43,44,45]. Also, the interference of the surface electromagnetic waves can be used in these purposes [46,47]. The latter mechanism can be used to create novel measuring and imaging systems on the basis of the method of the aperture synthesis [48]. For instance, using many-beam interference one can create the holographic beam splitters for the master oscillators in the complex powerful laser systems.

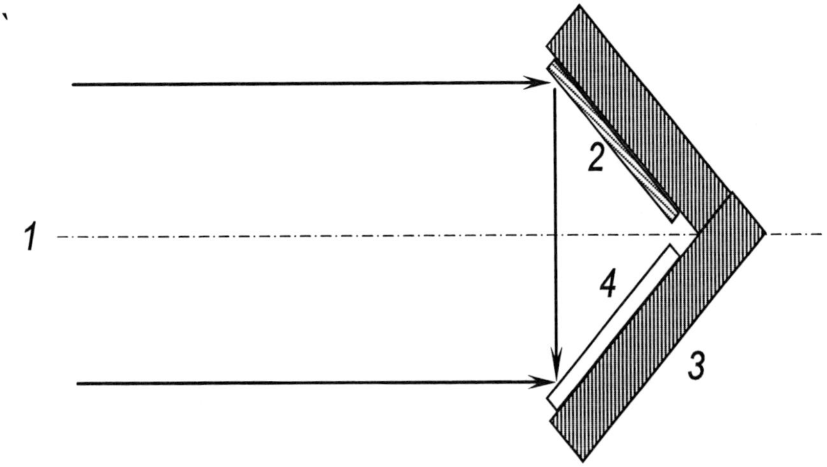

(a)

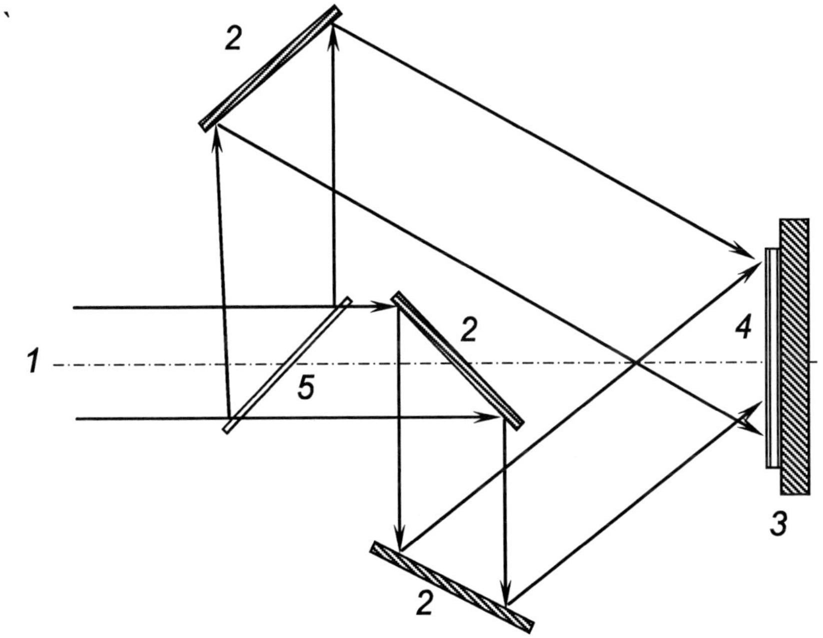

(b)

Figure 2. The schematic representations of the two experimental setups for Standing Wave Lithography: (a) — the Fresnel's mirror scheme; (b) — the interferometric scheme. *1* — an expanded laser beam; *2* — a high reflectivity mirror; *3* — the stage; *4* — the processed wafer; *5* — the semi-transparent mirror (the beam splitter).

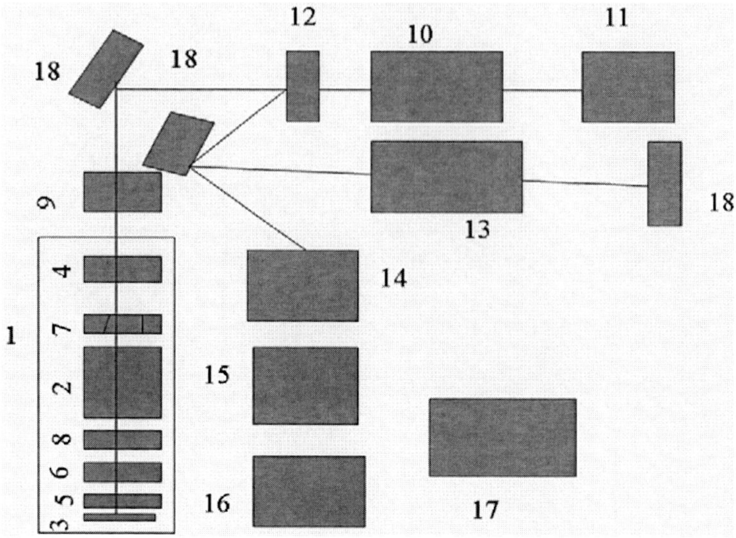

Figure 3. The schematic diagram of the laser system for four-beam standing wave XeCl pulsed excimer laser modification of the solid surfaces. *1* — the master oscillator (MO); *2* — the amplifying module; *3* and *4* — the back and the front mirrors of the cavity, respectively; *5, 6* — the Fabry-Perot etalons (FPs); *7* — the mode selection diaphragm; *8* — the polarizer; *9* — the beam expander; *10* — the first stage optical amplifier; *11* — the phase conjugation (PC) scheme; *12* — the beam decoupler; *13* — the second stage optical amplifier; *14* — the modified object; *15* — synchronization; *16* — the vacuum system; *17* — the scheme for measuring the spatial distribution of the laser radiation; and *18* — the mirrors.

In the experiments described in the present book, the interferometric scheme was used. It has employed a three-cascade UV XeCl-pulsed excimer laser designed and built at the Institute of Applied Physics, Russian Academy of Sciences (Nizhny Novgorod, Russia). The laser parameters were the following: the emission wavelength $\lambda \approx 308$ nm, the longitudinal coherence length > 30 cm, the laser radiation bandwidth ≈ 0.02 cm^{-1}, the pulse duration ranging from 2 to 45 ns, the pulse energy up to 100 mJ, the beam divergence was close to the diffraction limit. The schematic drawing of the experimental setup is presented in figure 3.

The laser system consisted of a master oscillator (MO) (1), of two cascades of the optical amplifiers, and of the optical elements for forming the laser beam(s) geometry and properties for matching between the amplification cascades. The master oscillator consisted of an amplifying module (2) with the pumping by a transverse discharge and automatic ionization of the active volume (33 × 2.8 × 0.7 cm^3 in volume). The cavity consisted of two planar mirrors with the reflection coefficients of 98% (3) and 50% (4).

In contrast to the conventional design, which utilizes the diffraction gratings in the master oscillator as the spectrum purifiers [49,50], two Fabry — Perot interferometers (FPs) (5, 6) with the sharpness ≈ 10 were employed were used for the laser spectrum rectification. This approach allowed improving the stability of the output radiation mode structure considerably. The FPs were placed inside the thermostats with the temperature stability ≈ 0.1°C at the ambient temperature variations within ≈ 3°C. The design of two FPs and thermostats implemented in the laser setup ensures the stability of the spectral characteristics of the laser radiation for about one week. After this time, the FPs required readjustment. A diaphragm of 1.5 mm in diameter (7) was used to select a single transverse mode in the cavity. An optical dielectric polarizer (8) was installed inside the cavity to select the radiation polarization. The total cavity length was ≈ 1.5 m.

A beam expander (a co-focal telescope, 9) with the spatial filter for increasing the beam diameter and decreasing the level of amplified spontaneous radiation appeared in the MO was placed between the MO and the mirror (18).

The first stage amplifier (10) had the active volume of 33 × 2.8 × 1.4 cm^3. The radiation passed around the amplifier twice. The second pass was provided by a phase-conjugation (PC) scheme (11) where the radiation was extracted due to rotation of the beam polarization plane by 90° (12). Compressed SF_6 under the pressure of ≈ 175 psi had been employed for PC.

The four-way second stage optical amplifier (13) had an active volume of 65 × 3.5 × 1.6 cm^3. The radiation was extracted by a stimulated Brillouin scattering (SBS) mirror. After the second stage amplifier, the radiation was directed into the optical scheme and split into the required number of beams. The beams were then joined together onto the surface to be modified (positioned vertically, 14). The optical scheme consisted of four dielectric mirrors in the case of the two beam interference and of 11 mirrors in the case of the four beam one. The dielectric mirrors with the reflectivity $R \approx 0.5$ were employed as the beam splitters and in order to direct the two resulting coherent beams onto the surface to be patterned the high reflectivity ($R \approx 0.99$) dielectric mirrors were used.

A high-voltage switch with a set of the cable delay lines (15) was used to synchronize the MO and the amplifiers. A vacuum system with one vacuum pump (16) was used in manufacturing and circulating the working-gas mixtures in the amplifiers.

The setup included also a high-resolution grating spectrograph, a calorimeter for measuring the pulse energy, several multi-channel oscilloscopes with the high-speed photo-detectors for recording the radiation kinetics, and a scheme for measuring the spatial distribution of the laser radiation (17). Also, the setup included a number of the dielectric mirrors (18).

The laser was initiated by feeding a high-voltage start-up pulse to the active-volume discharger of the master oscillator. The delay of the start-up of the first and the second stage amplifiers with respect to the MO depended on the rise time of the radiation in the MO and on the light propagation time from the MO to the corresponding amplifier. Usually, this delay was several tens of nanoseconds. The stability of the delay of the control pulses was ≈ 2 ns. The application of the thermally stabilized optics, of the spectral and spatial filtering, and of the SBS mirror made it possible to maintain a high quality of the laser beams during amplification. The SBS mirror was also used to shorten the laser pulses and to filter out the amplified spontaneous radiation. In most experiments the pulse duration was ≈ 7 ns unless other values were specified. All the experiments described below were performed in the ambient air at room temperature. Typically, the modified area on the sample surface was of several square millimeters.

The intensity pattern in space in the case of Two Beam Standing Wave Lithography in the symmetric scheme (i.e. when the angles of incidence of both interfetting laser beams are equal to ϑ) is given by [2]

$$\begin{aligned} I(x,z) &= E^2\left|\mathbf{e}_1\exp(ikx\sin\vartheta)+\mathbf{e}_2\exp(-ikx\sin\vartheta)\right|^2 = \\ &= E^2\left[2+\mathbf{e}_1\mathbf{e}_2^*\exp(i2kx\sin\vartheta)+\mathbf{e}_1^*\mathbf{e}_2\exp(-i2kx\sin\vartheta)\right]= \\ &= 2E^2\left[1+\sigma\cos(2kx\sin\vartheta)\right] \end{aligned} \tag{1}$$

where k = 2πn/λ, n is the refractive index of the medium (n = 1 for air), λ is the wavelength of the laser radiation, $\mathbf{e}_1$ and $\mathbf{e}_2$ are the unit polarization vectors for the two interfering electromagnetic waves. The period of the interference pattern along the x direction is $\lambda/2n\sin\vartheta$. Assuming $n = 1$ (in ambietn air or in vacuum), for example, at $\vartheta = 75°$, $\sin\vartheta = 0.97$, so the limiting period of the interference pattern is ≈ λ/2, which it is relatively easy to reach experimentally.

According to Snell's law, this period is invariant across any $\vartheta = \pi/2$ layer boundaries such as the lithography stack [2]. However; this is not true for the contrast. In addition, the contrast of the interference pattern depends on the polarization of the interfering laser beams [45]. Let us consider the following two cases: i) the polarization planes of both beams are perpendicular to the respective incidence planes (below referred to as TE—TE polarization configuration); and ii) the direction of polarization in one beam is perpendicular to the plane of incidence, while in another it is parallel to the plane of incidence (TE—TM configuration). The maximum contrast is achieved in the TE—TE case ($\sigma = 1$ in Eq.1). In the TE—TM configuration the contrast is reduced as a result of the π-phase shift between the $\mathbf{e}_1$ and $\mathbf{e}_2$ contributions to the interference intensity. In this case, $\sigma = \cos 2\vartheta$.

In the case of patterning of a photo-resist, it is important to account for the propagation angles inside the photosensitive layer in order to evaluate the polarization dependence of the contrast [2]. The contrast reduction for the TE—TM polarization configuration becomes essential for the steep propagation angles.

In the case of the four-beam interference, the general interference pattern is described by more than 20 parameters [2]. When one aims to create a non-modulated uniform periodic two-dimensional pattern, it is sufficient to consider the three simplified interference schemes characterized by a much smaller number of parameters:

(i) In the first scheme, two pairs of the beams lie in two different planes of incidence. These planes are perpendicular to each other, and the bisectors of the angles between the beams in each of the incidence planes are perpendicular to the sample surface.
(ii) The second scheme differs from the first one in that the bisector between the beams in one of the pairs is not perpendicular to the sample.
(iii) In the third scheme, the planes of incidence are not perpendicular to each other.

The mathematical procedure in describing the interference patterns in these cases is obvious and can be found, for instance, in [43]. Without going into the details of trigonometric calculations here, we shall mention several specific features of the interference of the four laser beams in the cases listed above. In the first scheme of interference, all the maxima have equal amplitudes and are separated by equal intervals. In the second and the third schemes, the amplitudes of intensity maxima in the plane of the sample and the separations between these maxima vary periodically.

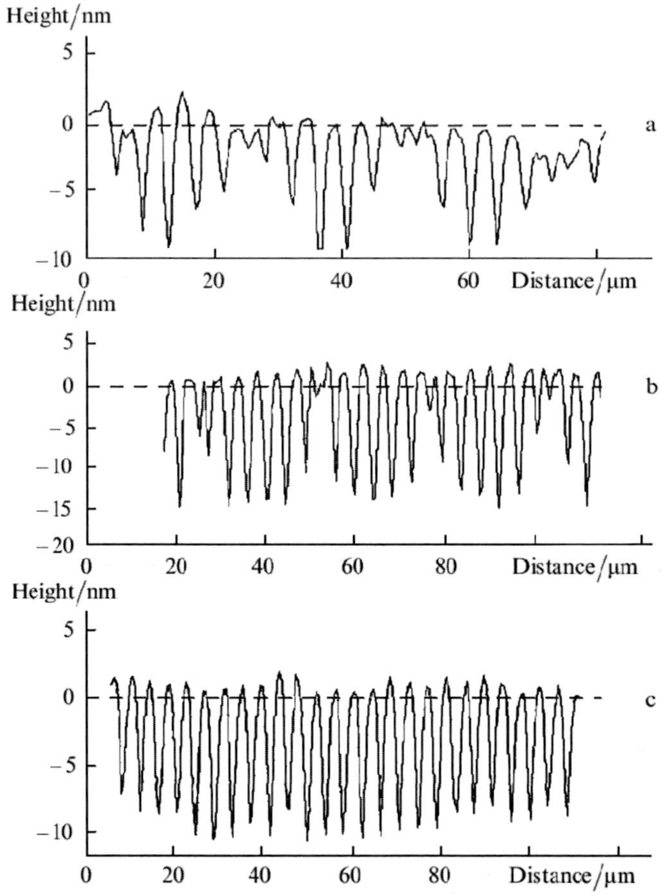

Figure 4. The effect of the deviation of the bisector of the angle between the two laser beams lying in the same plane of incidence from the normal to the sample surface on the relief of a Fe—C film patterned by direct sihlge-shot two-beam standing wave laser ablation: the bisector was not perpendicular to the sample surface (the deviation ≈ 2°) (a, b) and the bisector was almost perpendicular to the sample surface (the deviation ≈ 30`) (c). Reprinted from [6] under kind permission by *Turpion Co. Ltd.*

In all the above schemes, the interference pattern is highly sensitive to the polarizations of the interfering beams. Let us consider the following two cases: [45] i) the polarization planes of both pairs of the beams are perpendicular to the respective incidence planes (below referred to as TE—TE interference);

and ii) the direction of polarization in one pair of beams is perpendicular to the plane of incidence of these beams, while in the other pair of the beams it is parallel to the plane of incidence of these beams (TE—TM polarization configuration).

In [19] the pulsed laser modification of the surface of a Fe—C film with a thickness of several tens of nanometers deposited onto the Si substrates in the TE—TM four-beam interference polarization scheme was studied experimentally. The profiles of the modified Fe—C film surface are shown in figure 4. The profiles have been measured by Maxim GP 200 profilometer with ≈ 0.5 μm in-plane resolution and ≈ 0.2 nm vertical one. The profiles in figure 4 (a) and (b) correspond to the case when the bisector of one of the beam pairs is not perpendicular to the sample surface. In this case, the depth of surface modification varies across the sample. These variations became less expressed with decreasing angle between the bisector and the normal to the sample surface [figure 4 (c)]. From these experiments, one can conclude than in order to achieve the acceptable scatter of the intensity in the interference pattern maxima within 10%, the deviation of the angular beam alignment should be less than 1°.

It is worth noting that the Bragg mirrors with the spatially-modulated period can be formed also using the interference of the two laser beams with the different non-planar wave fronts [51]. The optical filters with complex transmission spectra can be created using the interference of two beams having several narrow lines in their emission spectra [52]. The same results can be obtained using the interference of three (or more) coherent monochromatic laser beans aligned in a single plane of incidence.

If all the interfering beams are inclined at the same angle to the surface normal, the resulting aerial image is again independent of the position of the wafer; however, if the beams are at different angles, the aerial image is no longer invariant along the direction and the placement of the wafer becomes important.

Most papers cited above were devoted to Standing Wave Laser Lithography where the interfering laser radiation was used to pattern a photo-resist followed by development and further processing as in standard UV lithography. In this book, we present mostly the results of forming the submicron periodic structures (including almost perfectly arranged periodic arrays of the holes in the metal films and of the semiconductor nanoislands) by direct Standing Wave Laser Modification using a powerful pulsed UV laser. Let us consider briefly the mechanisms of the solid surface modification in this process.

The interest to the effect of the laser radiation on the surface of various materials originates from the development of the laser technologies in the 1960-s. Many fundamental and applied problems have been considered since that time [38,53].

A powerful standing wave laser pulse results, first of all, in rapid (up to 10^{11} °C·s^{-1}) heating up of the patterned surface in the local nanometer sized areas. This heating may result in the material evaporation [54], in the re-crystallization of the films of various chemical composition [55,56,57], in the acceleration of the chemical reactions (in particular, in the laser-induced oxidation), in enhancement of the surface diffusion, etc. For example, this method was applied to the local nanodimensional modification of paramagnetic films with their conversion into the ferromagnetic state that is promising for magnetic data storage [58]. However, the effect of the laser radiation on a solid surface may not be limited to a pure thermal heating only. For example, it may result in a photochemical effect as well.

In order to understand these processes, it is important to know the temperature distribution in the structures subjected to the standing wave pulsed laser radiation. Knowing this distribution is also useful in studying the thresholds for the laser destruction of the metal coatings [59] and for excitation of the surface waves.

In the next section, a detailed two-dimensional calculation of the heat release in a metal film on a heat-conducting substrate [60] are presented. Then this theory will be applied to analysis of the experimental results presented in the following sections. It will be demonstrated that thermal physical properties both of the film and of the substrate play an important role when a standing wave laser radiation is employed to produce a submicron structures on the surface.

B. The Spatial Distribution of the Temperature in a Film on a Substrate Heated by a Surface Standing Wave of Laser Radiation

This section is devoted to an analysis of the heat transfer in a film of finite thickness on a semi-infinite substrate. Here we assume that the film surface is heated from the surface by a spatially periodic (in the film surface plane) heat source (namely, the standing wave of the laser radiation). This approximation is valid when the absorption length of the laser radiation employed $\beta(\lambda)^{-1} \ll L$

where β is the optical absorption coefficient of the modified material at the wavelength of λ and L is the film thickness. For $\lambda = 308$ nm this conditions is satisfied well for most metals (at reasonable L) and semiconductors. Here, we limited our analysis to considering the thermal effects only neglecting other effects of laser radiation on the film material listed in the previous subsection.

The schematic drawing of the configuration of the film on the substrate is shown in figure 5. The film has its boundaries at $x = 0$ and at $x = L$, and the substrate is located in the region $x < 0$. The thermal parameters of the film and of the substrate are indicated by the subscripts "1" and "2", respectively. Let us assume that the thermal flux q_{in} incident onto the film surface at $x = L$ is spatially-modulated along the y axis as

$$\begin{cases} q_{in}(y) = A + B\cos\left(\dfrac{2\pi y}{d}\right) \\ A \geq B \geq 0 \end{cases} \qquad (2)$$

where d is the period of the standing wave.

A two-dimensional heat conduction equation describing the spatial distribution of the temperature as a function of the time t $T(x, y, t)$ can be written as

$$\frac{\partial T_{1,2}}{\partial t} = \alpha_{1,2}\left(\frac{\partial^2 T_{1,2}}{\partial x^2} + \frac{\partial^2 T_{1,2}}{\partial y^2}\right), \qquad (3)$$

where $\alpha_{1,2} = k_{1,2}/\rho_{1,2}c_{1,2}$ are the temperature conductivities, $k_{1,2}$ are the heat conductivities, $\rho_{1,2}$ are the densities, and $c_{1,2}$ are the specific heat capacities for the film material and for the substrate one, respectively.

The initial condition for the problem considered here is that the temperatures in both the film and the substrate are constant at $t = 0$:

$$T_{1,2}(x, y, 0) = T_0. \qquad (4)$$

The boundary conditions are written as follows. For $x = L$, the heat flux entering the film is

$$k_1 \frac{\partial T_1}{\partial x} = q_{in}. \qquad (5)$$

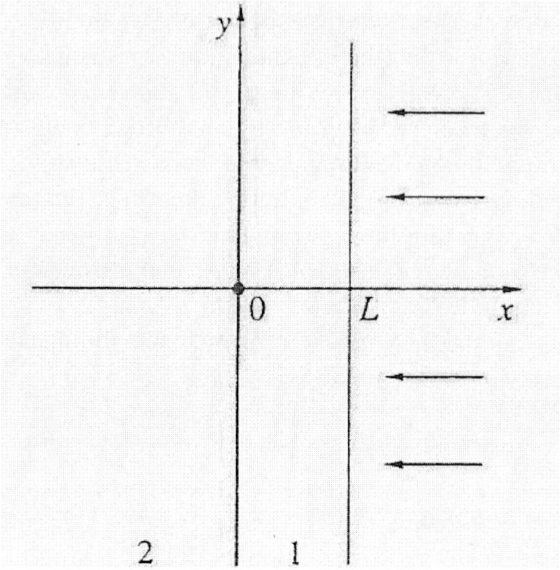

Figure 5. A schematic drawing of a film of the thickness L on a semi-infinite substrate. Region (1) is the film and region (2) is the substrate. The direction of the heat flux incident onto the film surface at $x = L$ is indicated by the arrows.

For $x = 0$, the continuity condition for the heat flux

$$k_1 \frac{\partial T_1}{\partial x} = k_2 \frac{\partial T_2}{\partial x} \tag{6}$$

and the condition of equal temperatures

$$T_1(0, y, t) = T_2(0, y, t) \tag{7}$$

should be satisfied. For $x \to -\infty$, the temperature of the substrate should tend to T_0:

$$T_2(x, y, t)_{x \to -\infty} \to T_0 . \tag{8}$$

Under these conditions, the solution of Eq. (3) is

$$T_{1,2}(x, y, t) = T_{1,2}^{c}(x, t) + T_{1,2}^{a}(x, t) \cos\left(\frac{2\pi y}{d}\right) \tag{9}$$

Applying the Laplace transformation with respect to the time t to Eq. (3) and to the boundary conditions (5) — (8), one can obtain the equations for the Laplace images $T^{c}{}_{1,2}(x, p)$ and $T^{a}{}_{1,2}(x, p)$.

The solutions of Eq. (3) have different forms in the regions 1 and 2. Here we consider the solution for the temperature distribution in the film ($0 < x < L$) only:

$$T_1(x, y, t) = T_1^c(x,t) + T_1^a(x,t)\cos\left(\frac{2\pi y}{d}\right). \tag{10}$$

The constant component $T_1{}^c$ is given by the following expression:

$$T_1^c(x,t) = T_0 + \frac{A}{k_1}\sum_{n=0}^{\infty}\gamma^n\left\{2\sqrt{\frac{\alpha_1 t}{\pi}}\left[\exp\left\{\frac{[(2n+1)L-x]^2}{4\alpha_1 t}\right\} + \gamma\exp\left\{\frac{[(2n-1)L+x]^2}{4\alpha_1 t}\right\}\right]-\right. \tag{11}$$
$$\left.-[(2n+1)L-x]\mathrm{erfc}\left[\frac{(2n+1)L-x}{2\sqrt{\alpha_1 t}}\right] - \gamma[(2n+1)L+x]\mathrm{erfc}\left[\frac{(2n+1)L+x}{2\sqrt{\alpha_1 t}}\right]\right\}$$

where

$$\gamma = \frac{1-\sigma}{1+\sigma}, \tag{12}$$

and

$$\sigma = \sqrt{\frac{k_2\rho_2 c_2}{k_1\rho_1 c_1}}. \tag{13}$$

The amplitude of the oscillating component of the temperature distribution along the y axis $T_1{}^a$ is given by

$$T_1^a(x,t) = \frac{B}{k_1}(D - I - Q) \tag{14}$$

where

$$D=\frac{d}{2\pi}\cdot\frac{\cosh\left(\frac{2\pi x}{d}\right)+\delta\sinh\left(\frac{2\pi x}{d}\right)}{\sinh\left(\frac{2\pi L}{d}\right)+\delta\cosh\left(\frac{2\pi L}{d}\right)} \tag{15}$$

and $\delta = k_2/k_1$. The form of the functions I and Q depends on the relationship between the thermal conductivities α_1 and α_2. This effect is related to the fact that the number of poles and branching points of the function $T_1^a(x,p)$ depends on the above relationship. When $\alpha_1 > \alpha_2$

$$I=\frac{2\delta L}{\pi}\int_0^\infty \exp\left\{-\frac{\alpha_1 t}{L^2}\left[\left(\frac{2\pi L}{d}\right)^2+u^2\right]\right\}\beta(u)\cos\left[u\left(1-\frac{x}{L}\right)\right]\times$$
$$\times\frac{u\partial u}{\left[\left(\frac{2\pi L}{d}\right)^2+u^2\right]\left(u^2\sin^2 u+\delta^2\beta^2(u)\cos^2 u\right)} \tag{16}$$

$$Q=\frac{2\delta L}{\pi}\int_0^\varepsilon \exp\left\{-\frac{\alpha_1 t}{L^2}\left[\left(\frac{2\pi L}{d}\right)^2-u^2\right]\right\}\tilde{\beta}(u)\cosh\left[u\left(1-\frac{x}{L}\right)\right]\times$$
$$\times\frac{u\partial u}{\left[\left(\frac{2\pi L}{d}\right)^2-u^2\right]\left(u^2\sinh^2 u+\delta^2\tilde{\beta}^2(u)\cosh^2 u\right)} \tag{17}$$

where

$$\varepsilon=\frac{2\pi L}{d}\sqrt{1-\frac{\alpha_2}{\alpha_1}}, \tag{18}$$

$$\beta(u)=\sqrt{\left(\frac{\alpha_1}{\alpha_2}-1\right)\left(\frac{2\pi L}{d}\right)^2+\frac{\alpha_1}{\alpha_2}u^2}, \tag{19}$$

and

$$\tilde{\beta}(u)=\sqrt{\left(\frac{\alpha_1}{\alpha_2}-1\right)\left(\frac{2\pi L}{d}\right)^2-\frac{\alpha_1}{\alpha_2}u^2}\,. \tag{20}$$

When $\alpha_1 < \alpha_2$, the following expressions can be obtained:

$$I=\frac{2\delta L}{\pi}\int_{\omega}^{\infty}\exp\left\{-\frac{\alpha_1 t}{L^2}\left[\left(\frac{2\pi L}{d}\right)^2+u^2\right]\right\}\beta(u)\cos\left[u\left(1-\frac{x}{L}\right)\right]\times$$
$$\times\frac{u\partial u}{\left[\left(\frac{2\pi L}{d}\right)^2+u^2\right]\left(u^2\sin^2 u+\delta^2\beta^2(u)\cos^2 u\right)} \tag{21}$$

where

$$\omega=\frac{2\pi L}{d}\sqrt{\frac{\alpha_1}{\alpha_2}-1}\,, \tag{22}$$

and

$$Q=\sum_i\frac{\exp(-\rho_i t)}{C_i}\left[r_i\cos(r_i x)+\delta l_i\sin(r_i x)\right], \tag{23}$$

$$r_i=\sqrt{\frac{\rho_i}{\alpha_1}-\left(\frac{2\pi}{d}\right)^2}\,, \tag{24}$$

$$l_i=\sqrt{\left(\frac{2\pi}{d}\right)^2-\frac{\rho_i}{\alpha_2}}\,, \tag{25}$$

and ρ_i are the roots of the equation

$$\sqrt{\frac{\rho}{\alpha_1}-\left(\frac{2\pi}{d}\right)^2}\,tg\left[L\sqrt{\frac{\rho}{\alpha_1}-\left(\frac{2\pi}{d}\right)^2}\right]=\delta\sqrt{\left(\frac{2\pi}{d}\right)^2-\frac{\rho}{\alpha_2}}\,. \tag{26}$$

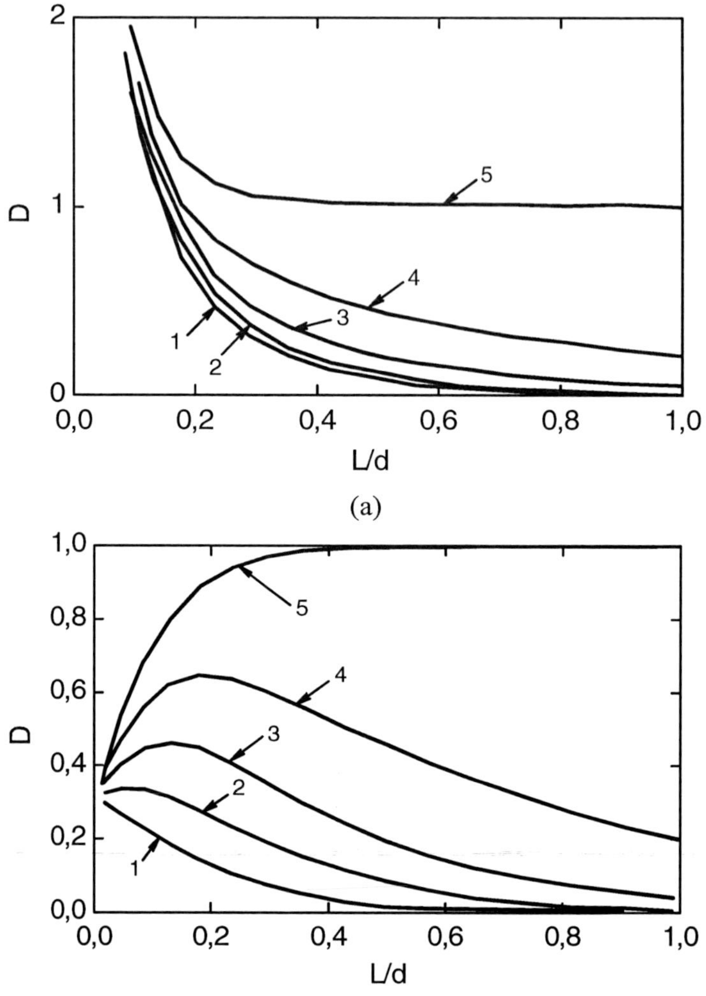

Figure 6. The effect of the metal film thickness L and of the period of the spatial modulation of the heat flux d on the steady-state amplitude of the temperature distribution in the y direction for various depths x. The substrate material: (a) — silica glass; (b) — silicon. x/L: *1* — 0, *2* — 1/4, *3* — 1/2, *4* — 3/4, and *5* — 1.

All the roots ρ_i of Eq. (26) should be determined, and the summation in Eq. (23) should be performed over the whole range of i. The roots ρ_i fall into the range from $\rho = \alpha_1(2\pi/d)^2$ to $\rho = \alpha_2(2\pi/d)^2$. The total number of roots of Eq. (26) is $[z] + 1$, where $[z]$ is the integer part of $2L/d(\alpha_2/\alpha_1 - 1)^{0.5}$.

The solution $T_2(x, y, t)$ for the substrate that is not written here also satisfies Eq. (3). The boundary conditions (4) — (8) are also satisfied by the solutions found. We failed to verify the fulfillment of the initial condition $T_{1,2}(x, y, 0) = T_0$ analytically, but the computer calculation demonstrated that this condition was satisfied as well. If one considers the condition (4) to be satisfied, the improper integrals in (16) and (18) for $t = 0$ can be expressed in terms of function (15) and, correspondingly, in terms of (17) and (19).

It should be noted that the integrals I and Q entering expression (14) for the amplitudes $T_1^a(x,t)$ tend to zero at $t \to \infty$. This means that $T_1^a(x,\infty) = BD/k_1$ is a steady-state value of the amplitude of the temperature distribution component oscillating with varying y.

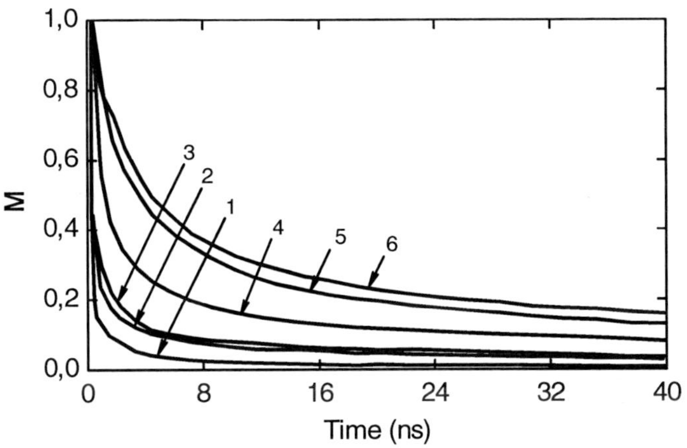

Figure 7. The depth of the spatial modulation of the temperature M with the period d for a Fe film of thickness $L = 20$ nm on silica glass (*1* and *5*), Fe (*2* and *6*), and Si (*3* and *4*) substrates as a function of the heating time t. The spatial modulation period of the heating d =200 nm (*1, 2,* and *3*) and d = 800 nm (*4, 5,* and *6*).

Figure 6(a) and (b) shows $\overline{D} = 2\pi D/d$ as a function of L/d for various values of depths x normalized to the film thickness L for a metal film on the silica glass substrate and on the Si one, respectively. These represent the behavior of the steady-state value of the temperature distribution component varying along y direction for the metal films on various substrates. A dramatic effect of the thermal properties of the substrate on the spatial distribution of the film temperature is seen clearly. In the case of the silicon substrate $k_1 < k_2$

and D decreases with decreasing L/d starting from a certain value of L/d. In this case, $\overline{D} \rightarrow k_1 k_2$. For the silica glass substrate, $k_1 > k_2$ and D increases with decreasing L/d.

One can see from Eqs. (16) — (21) that the settling time for the steady-state value of $T_1^a(x,t)$ is determined by the two parameters, namely by $t_L = L^2/\alpha_1$ and by $t_d = d^2/(2\pi)^2\alpha_1$. The effect of the latter parameter t_d is related to the presence of an exponential factor that does not enter the integrand. The effect of t_L entering the integrand, which is a function of four thermal physical parameters and of the quantity $2\pi L/d$, is more complicated. As follows from computer simulation, the settling time for the steady-state value of $T_1^a(x,t)(t_{st})$ for the metal film on the silica glass substrate decreased from $5t_d$ down to $2t_d$ as L/d increased from 0.025 to 0.25. For the silicon substrate, a different behavior of t_{st} was obtained. Namely, t_{st} increased from $0.3t_d$ up to $\approx t_d$ as L/d increased within the limits indicated above.

For certain applications, the parameter $M = T_1^a(x,t)/\left[T_1^c(x,t) - T_0\right]$ characterizing the depth of spatial modulation of the temperature is important. For the boundary conditions (4) — (8), $T_1^c(x,t) \rightarrow \infty$ for $t \rightarrow \infty$, and, consequently, $M \rightarrow 0$ for $t \rightarrow \infty$. This means that in order to obtain a desired value of M the duration of heating should be limited. The effect of the substrate parameters on $M(L, t)$ for various values of L and d is shown in figure 7 and figure 8. Note that the relative effect of the substrate is more pronounced as d decreases (figure 7).

For d = 200, 400, and 800 nm the parameter t_d is 0.085, 0.34, and 1.35 ns, respectively. As follows from figure 5 and figure 6, the modulation depth is 15-20% for the heating time $t < 10t_d$.

The main conclusions from the analysis and computer simulations, results of which are presented in this section, are as follows:

(i) The parameter L/d should be smaller than 0.1 for the heating to be sufficiently uniform over the film thickness.

(ii) The heating time t should be less than $10t_d$ in order to provide the modulation depth of the film temperature (in the y direction) greater than 15%.

(iii) When the structures with the period $d < 500$ nm are produced, the modulation depth M for the Si substrates is 2 to 4 times greater than for the silica glass ones. The effect is attributed to higher heat conductivity of Si compared to the one of silica glass.

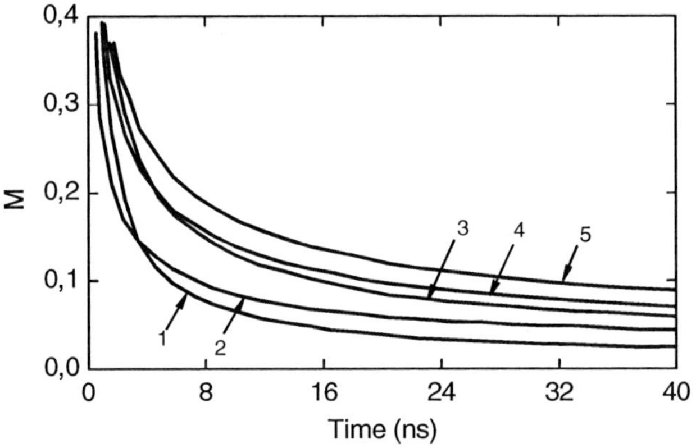

Figure 8. The depth of the spatial modulation of the temperature M for a Fe film on a silica glass substrate (1, 2) and on a Si one (3, 4, 5) as a function of the heating time t. The period of the spatial modulation of the heating source $d = 400$ nm. The film thickness L, nm: *1, 4* — 40, *2, 3* — 10, and *5* — 200.

Finally, it should be noted that the problem considered here can be extended to the case when the input flux q_{in} is an arbitrary function of y. This can be done by representing it in terms of a Fourier integral and carrying out inverse transformation of Fourier images obtained from the solutions of the heat conduction equations (14) — (26). The time dependence of the input heat flux q_{in} can also be included with the use of Duhamel's integral.

Chapter 3

Forming the One- and Two-Dimensional Periodic Relief on the Fused Silica and Si Surfaces by Standing Wave -Pulsed Laser Lithography

A. One-dimensional Periodic Nanorelief on the Fused Silica and Si: the Phase Masks and the Antireflection Structures

In this section, the results on forming the one-dimensional gratings on the surface of the fused silica substrates and on the Si ones using Two-Beam Standing Wave UV Pulsed Laser Lithography are reviewed. These one-dimensional gratings were intended for application as the phase masks and as the antireflection structures for the visible spectral band and for the near IR one, respectively. In these experiments, we applied the traditional Standing Wave Lithography technique including exposing the photo-resist films on the surface of the fused silica substrates and of the Si ones to the standing wave UV-pulsed laser radiation followed by a standard lithographic processing including the development of the exposed resist and by dry etching. One of the most promising applications of this technology is fiber and integrated optics, namely forming the phase masks and Bragg mirrors with controlled reflection/transmission spectra. Such gratings made from Si are promising for various applications in the IR wavelength band, for instance, in the solar cells as the antireflection coatings [61] and in the integrated optics, especially, in Si based photonics, which is a new field of science and technology developed rapidly in recent years [62]. There the regular surface relief can be applied as the Bragg mirrors in the Si based planar and ridge waveguides (in the photodetractors, multiplexors, etc.), in the distributed feedback (DFB) elements of the Si-based lasers, in the one-dimensional photonic crystals, etc.

For forming the phase masks, E-Beam Lithography is used the most widely. The studies presented in this section were devoted to development of a simple and cost-effective method of Standing Wave Lithography using a powerful XeCl excimer pulsed laser for forming the one-dimension gratings. One of the most important advantages of the Standing Wave Pulsed Laser Lithography is much less effect of the vibration environment on the nanoscale pattern quality. Besides, using a powerful laser allows increasing the pulse energy up to several tens of joules that in turn would allow exposing relatively large areas (up to 1000 cm^2) at a single shot. On the other hand, the breakdown threshold for fused silica exposed to the UV standing wave laser radiation with $\lambda = 308$ nm is ≈ 15 $J{\cdot}cm^{-2}$. Such a high value does not allow the formation of a regular periodic relief over an area greater than ~ 1 cm^2 by direct Standing Wave Laser Ablation.

Another reason for using a resist instead of direct standing wave laser ablation was a high enough aspect ratio of the surface corrugation required for the phase masks and the antireflection gratings, which could hardly be achieved using the direct ablation. Let us estimate the parameters of the surface nanorelief required for the efficient phase masks and antireflection gratings. For the phase masks, the condition of minimum transmission in the zero order of diffraction (MTZOD) is an obligatory one. The quality of the phase mask can be considered to be satisfactory when MTZOD is less than 10%. From the elementary physics considerations, MTZOD is obtained when the corrugation depth $h \approx \lambda/2(n - 1)$ where λ is the wavelength of the incident light used for the pattern transfer and n is the refractive index of the phase mask material. This condition provides MTZOD at normal incidence of the light onto the phase mask with a rectangular relief. However, this estimate is quite rough; and modern technological developments require more accurate theory as well as the experimental verification of the latter.

For the antireflection gratings, the following two conditions are to be satisfied. First, the period of the surface relief d should be less than λ_a/n where λ_a is the wavelength of the light, which the antireflection grating is designed for. Second, the corrugation height h should equal approximately to $\lambda_a/4$. The first condition is required in order to suppress the diffraction of the higher orders inside the antireflection grating material while the second one provides the MTZOD at the normal incidence.

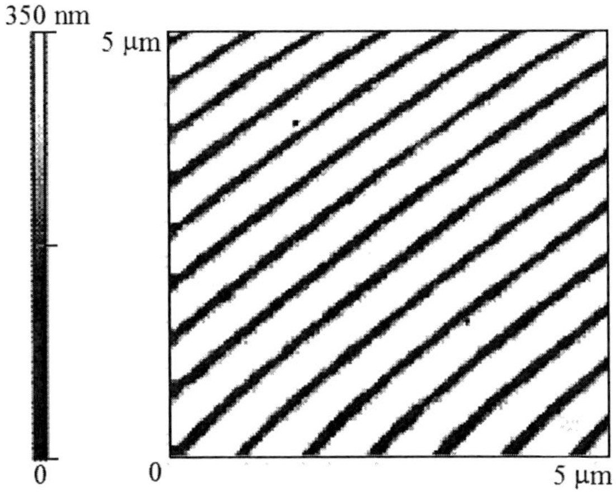

(a)

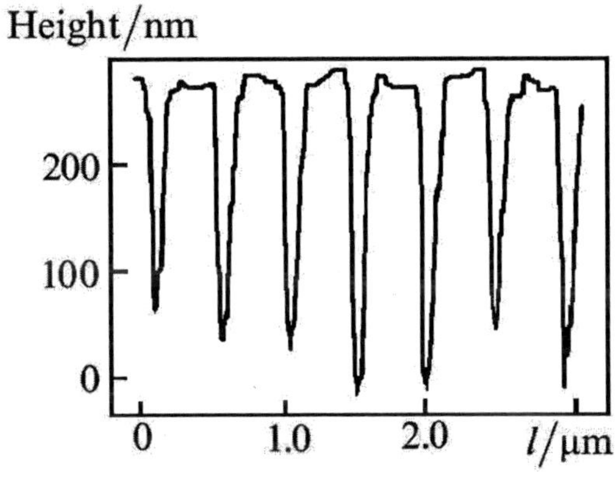

(b)

Figure 9. An AFM image (a) and a line profile (b) of a phase mask on the surface of a silica glass substrate formed by Standing Wave Excimer XeCl Pulsed Laser Lithography with reactive ion etching. Reprinted from [16] under kind permission by *Turpion Co. Ltd.*

The forming of the relief on the surfaces of the silica glass substrates was performed using a double layered resist mask consisting of a Cu film of ≈ 50 nm in thickness deposited onto a ≈ 400 nm thick photo resist film. We have used the two-beam XeCl laser interference scheme with the energy densities ≈ 30 mJ/cm^2. The exposed resist was developed in an alkaline water solution and Cu film was etched through using Ar^+ ion beam etching. The silica glass was etched using Reactive Ion Beam Etching in a flux of CF_4 — O_2 gas mixture. The etching rates were ≈ 30 nm/min for fused silica and for the photo-resist and ≈ 3 nm/min for Cu. This way, the relief on the surface of the fused silica substrate up to ≈ 500 nm in depth has been formed.

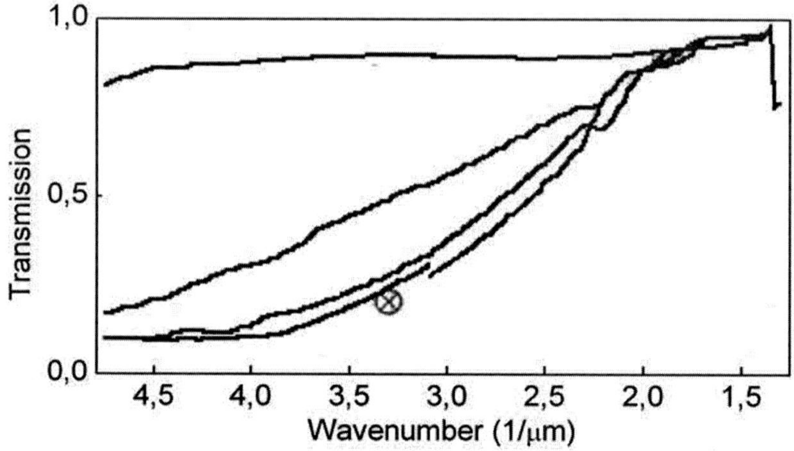

Figure 10. The optical transmission spectra of the reference sample (the initial fused silica substrate) (1) and of the phase mask formed by Standing Wave Excimer XeCl Pulsed Laser Lithography with reactive ion etching, the morphology of which is presented in figure 13 (2). ⊗ — the result of the measurement with XeCl laser as a probing light source. Reprinted from [16] under kind permission by *Turpion Co. Ltd.*

In figure 8 (a) and (b) an AFM image and a line profile, respectively of a phase mask formed on a fused silica substrate by the process described above are presented. The period of the grating d was ≈ 460 nm and the averaged groove depth h was ≈ 280 nm. It should be noted that the width of the grooves and the one of the ridges are not equal to each other. These parameters depend on the conditions of the developing of the photo-resist, as well as of the ones of the etching of the metal mask and of the silica glass substrate. In order to optimize the phase mask relief, further experiments are necessary.

The optical transmission measurements in the zero order of diffraction have been carried out using Specord M40 Carl Zeiss Jena spectrophotometer. Also, a special measurement has been carried our using XeCl laser as a probing light source. In the latter case, the beam diameter was expanded to ≈ 5 mm. In order to increase the measurement accuracy, a double beam technique with a reference beam has been applied. The results of the optical transmission measurements are presented in figure 10. Curve 1 represents the optical transmission spectrum of a flat fused silica substrate served as a reference sample. Curve 2 represents the transmission spectrum of the sample the AFM image of which was shown in figure 9. The minimum TCZOD took place at $\lambda \approx 250$ nm and the antireflection properties were observed at $\lambda_a > 630$ nm. So far, the structure obtained can serve both as a phase mask at $\lambda \approx 250$ nm and as an antireflection grating at $\lambda > 630$ nm at the same time.

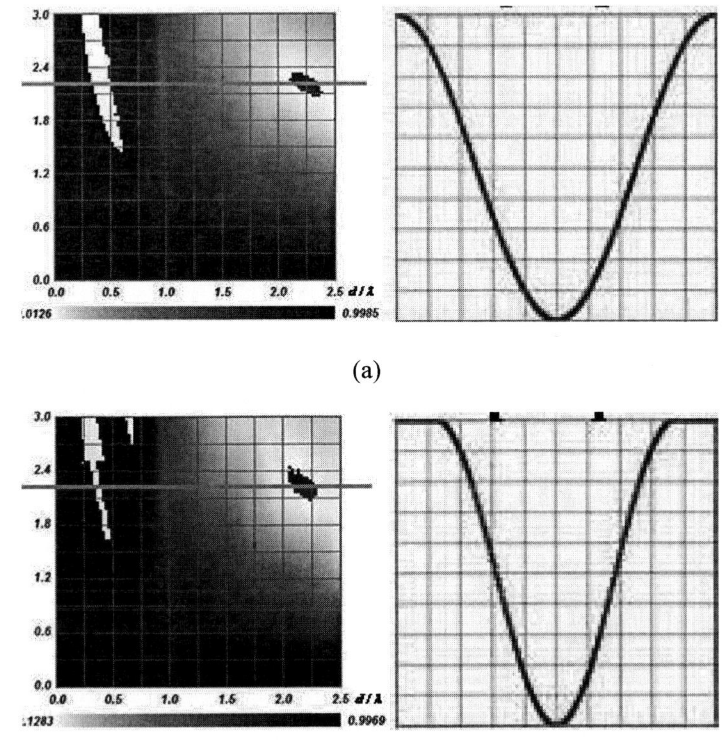

(b)

Figure 11. The calculated grayscale diagrams of the optical transmission of the goffered-fused silica substrate for the sinusoidal model surface profile (a) and for the truncated sinusoidal one (b).The fields close to the maximum and the minimum transmissions are highlighted by the inverted colors. Reprinted from [16] under kind permission by *Turpion Co. Ltd.*

The transmittance of a goffered-fused silica surface has been calculated by the numerical solving of the integral equation for diffraction of a plane wave on a goffered surface [63]. The calculations were carried out for the two corrugation profiles: i) for the sinusoidal one [figure 11 (a)] and ii) for a truncated sinusoide [figure 11 (b)]. The latter was selected as a model profile to approximate the measured one presented in figure 9 (b). The refractive index of fused silica n was assumed to be 1.46 over the whole wavelength band, which the calculations were carried out for. The optical absorption in fused silica was neglected. The refractive index of the ambient medium was assumed to be equal to unity.

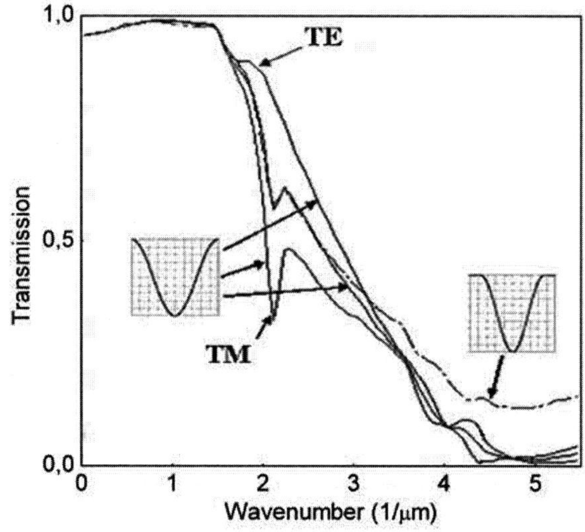

Figure 12. The calculated optical transmission of a goffered-fused silica surface versus d/λ ratio for the relief heights providing the minimum transmission (corresponding to the gray lines in figure 12). Reprinted from [16] under kind permission by *Turpion Co. Ltd.*

The calculated dependencies of the power transmittances T (averaged over the TE and TM polarization modes of the incident light) on the dimensionless parameters of the phase mask h/d and d/λ are presented in figure 11 as the grayscale maps for the respective corrugation profiles presented nearby. The fields of the near-maximum transmittance and of the near-minimum one are denoted by the inverted grayscale palette tones for clarity. In both cases of the sinusoidal surface profile and of the profile close to the one obtained in the experiment, the calculated values of T were close to unity within a rather wide range of the model parameters $\lambda > d$, $h > \lambda$, and $h > d/2$. The general pattern of the T(h/d, d/λ) chart depends on the exact shape of the surface profile weakly. For the ideal sinusoidal profile T reaches its minimum (T $\rightarrow$ 0) when $d/\lambda \approx$ 2.25 and $\pi h/d \approx 2.21$. For the actual corrugation profile minimum T is greater than 0,13 and its position in the h/d — d/λ axes is shifted to lower values of d/λ as compared to the case of the ideal sinusoidal profile.

In figure 12, the calculated profiles of T(d/λ) along the lines shown in figure 11 (for the values of h providing the minimum transmission) are presented. For the structure studied in the experiment $d \approx 460$ nm and $h \approx 320$ nm (i. e. $\pi h/d \approx 2.21$). Curve 1 (the bold line) presents T vs d/λ for the sinusoidal profile and curve 2 (the dashed bold line) — for the truncated sinusoidal one. These two curves represent the optical transmission spectra averaged over the TE and the TM modes to simulate the non-polarized natural light transmission. The respective solid curves represent the results of the calculations for the TE and the TM modes, respectively for the in-deal sinusoidal profile. The difference between these two curves is the most pronounced near $\lambda_1 \approx d$, which is the cut-off wavelength for the first diffraction maximum in vacuum. There is a narrow dip in the T(d/λ) curve at λ_1 for the TE mode while a smoothed step takes place for the TM one. It should be noted that the dip is seen in the averaged T(d/λ) curve at λ_1 as well, although the dip is less pronounced.

A rapid fall of T with decreasing λ takes place at transition from the longer wavelengths where T $\rightarrow$ 1 to the shorter ones. Another critical point important for the experimental data analysis is $\lambda_2 = nd$, which corresponds to the cut-off for the first harmonic in the substrate material. The calculations predict a step or even a peak in the T(d/λ) curves near λ_2 for both surface profile shapes considered above and for both incident light-wave polarizations.

Having measured λ_2, one can determine n if the period of the grating is known. For the sample studied in the present work, $\lambda^*_2 = dn \approx 650$ nm, $d \approx 460$ nm, and $n \approx 1.46$. This value was used in the calculations.

In figure 10 besides the measured optical transmission spectra of the goffered fused silica substrate the calculated ones are presented (curves 3 and 4). The latter were calculated by numerical solving of the integral equation for diffraction of light on a fused silica plate of finite thickness with a goffered front side taking into account the parasitic reflection of the transmitted light from the flat back side. The model optical transmission spectra were calculated for the truncated sinusoidal surface profile approximating the one of the experimental sample. The behavior of the measured curve indicates the domination of the TM-polarized radiation in the probing light since the dip near $\lambda = d$ featuring the TE-polarized light is not expressed. Fitting of the measured transmission spectrum by the calculated one revealed the TE/TM ratio in the probing light to be $\approx 1:4$.

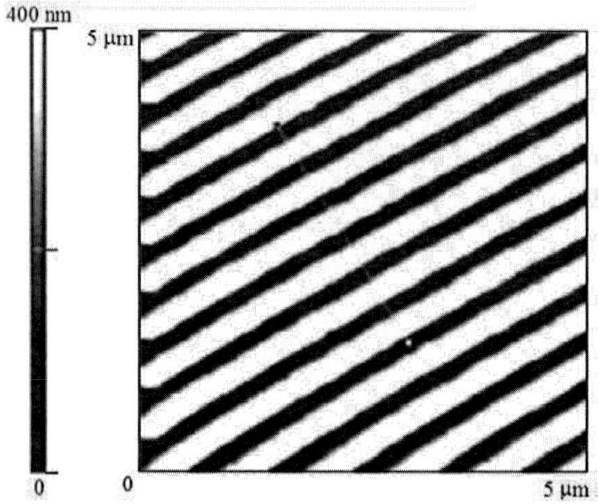

(a)

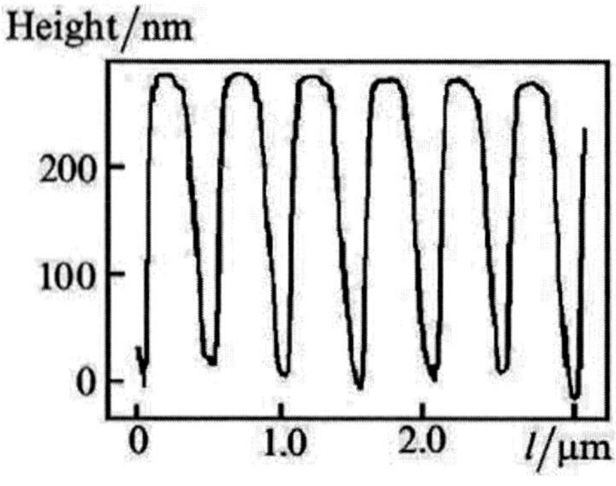

(b)

Figure 13. An AFM image (a) and a surface profile (b) of a one-dimensional grating on a Si single crystal substrate formed by Two-Beam Standing Wave Pulsed Laser Lithography using XeCl excimer laser to pattern the photoresist film on Si and subsequent Ar^+ ion etching. Reprinted from [16] under kind permission by *Turpion Co. Ltd.*

Curve 3 was calculated for $h = 280$ nm and curve 4 — for $h = 390$ nm. Curve 3 is consistent with the experimental transmission spectrum (curve 2) within the longer wavelength band however at shorter wavelengths the calculated transmission was greater than the measured one. On the contrary, curve 4 concedes with the measured curve 2 better as shorter wavelengths and worse at longer ones. This disagreement could be attributed to not accurate enough approximation of the actual surface profile by the model one (the truncated sinusoid). Also, the scatter of the groove depth and width was not taken into account in the calculations.

In figure 13 (a) and (b) an AFM image and a surface profile of a grating formed on the surface of a single crystal Si wafer are presented. The patterned area was ≈ 6 mm in diameter. The period of the grating d was ≈ 460 nm and the groove depth h was 350 – 370 nm. Using the formulae given above, one can estimate the cut-off wavelength for the antireflection band to be $\lambda_1 \approx 1.5$ μm.

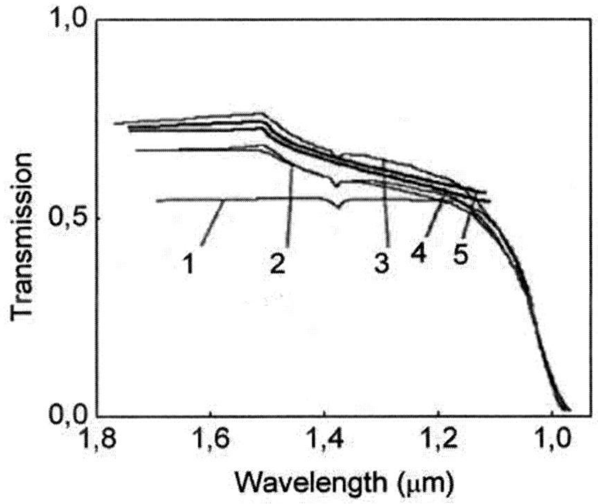

Figure 14. The measured optical transmission spectra of the flat Si wafer (1) and the ones for the goffered Si wafers (2 — 4); the calculated optical transmission spectra for the sinusoidal model surface profiles with $h = 350$ nm (5) and with $h = 400$ nm (6). Reprinted from [16] under kind permission by *Turpion Co. Ltd.*

In figure 14 the IR transmission spectra of the Si wafers measured using LOMO SP-20 spectrophotometer and the calculated ones are presented. Curve 1 presents the measured transmission spectrum of a flat Si wafer. Curves 2 and 3 are the measured transmission spectra of the patterned Si wafers with $h \approx$

350 nm and $h \approx 400$ nm, respectively. A significant increase in the IR optical transmission at $\lambda > 1,5$ μm has been observed in curves 2 and 3 that is in a good agreement with the rough estimate given above. More detailed analysis of the measured transmission spectra revealed the cut-off wavelength of the first harmonic in the substrate $\lambda_2 = nd \approx 1.53$ μm. From $d \approx 466$ nm [see figure 13 (b)] one can estimate $n \approx 3.28$.

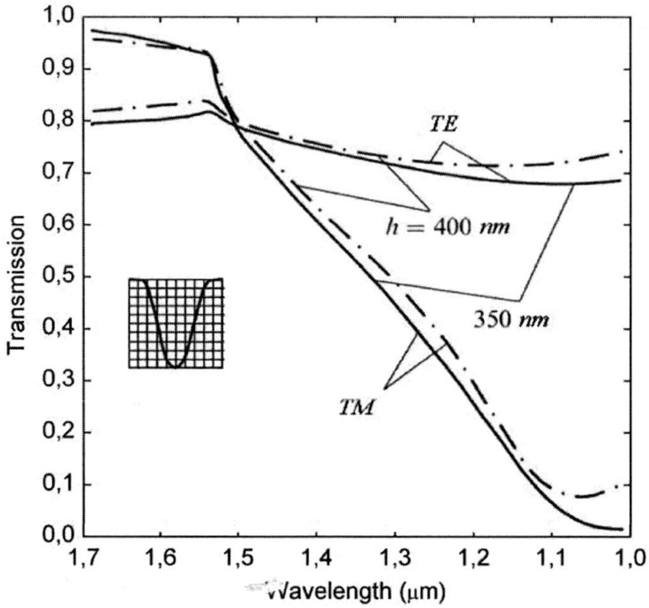

Figure 15. The calculated optical transmission spectra of a goffered Si surface for the model surface profile shown in the inset for different polarization modes. *h,* nm: *1* — 350; *2* — 400. Reprinted from [16] under kind permission by *Turpion Co. Ltd.*

The model transmission spectra (curves 4 and 5 in figure 14) were calculated just for this particular value of n and for the relief profile close to the one presented in figure 13 (b). The latter was approximated by a truncated sinusoide with the length of the plateaus on the tops equal to $d/5$ (as shown in the inset in figure 15) for the two values of h = 350 and h = 400 nm. The calculated IR transmission spectra were the weighted average of the ones for the TE and the TM modes (5 TE : 1 TM) as in the probing light used in the IR transmission measurements. In figure 15, the model IR transmission spectra for the TE and the TM modes calculated for the two values of h = 350 nm and

$h = 400$ nm are presented. The calculations were performed for $\lambda > 1.15$ μm where the measured transmission spectrum of the flat Si wafer was almost constant (see figure 14, curve 1). The optical absorption in Si was not taken into account since Si is almost transparent in the wavelength band of interest. The calculations demonstrated the IR transmission spectra to be different for the TM and TE modes essentially. On the other hand, a little difference was found between the transmission spectra calculated for $h = 350$ nm and for $h = 400$ nm.

In conclusion, a promising technology for forming the one-dimensional submicron periodic relief on fused silica substrates and on the Si ones using Standing Wave-Pulsed Laser Lithography have been developed. Comparing the experimental results on the optical properties of the test samples with the results of computer simulations demonstrates applicability of the structures made by the above technology as the phase masks for the fiber and integrated optics and as the antireflection structures for various photonic device applications.

B. Formation of the Two-dimensional Periodic Arrays of the Nanometer-Sized Pits with High-Aspect Ratio in Fused Silica

In this section, we present the experimental results on the forming of the two-dimensional submicron periodic relief on the fused silica substrates by Four-Beam Standing Wave Pulsed Laser Lithography using the four beam standing wave XeCl laser radiation to expose an UV photo resist film on the fused silica surface followed by development and wet or Ar^+ ion etching. The periodic arrays of the nanometer-sized pits in the fused silica surface have been obtained.

The nanometer scale periodic patterning of the fused silica surface is an important applied problem. The relevant technology could be applied to the formation of various integrated optical and electronic devices. Its potential applications include, first of all, the phase masks for forming the Bragg mirrors for the planar optical waveguides [64,65,66], the planar photonic crystals, the punches for the nanoimprint technology [67,68,69] the broadband antireflection structures [70,71], etc. [72,73]. From the technological viewpoint, the direct patterning of fused silica surface by Four Beam Standing Wave Laser Ablation would be extremely promising. However, the relatively large value of the breakdown threshold for fused silica at $\lambda \approx 308$ nm (≈ 15 $J{\cdot}cm^{-2}$, as had already been mentioned above) would make us to reduce the patterned area downto ~ 1 mm^2 (at the maximum pulse energy of the laser available ≈ 300 mJ) and therefore to loose in the main advantage of the direct standing wave laser modification namely, in the productivity. So far, in order to make a periodic pattern on the fused silica surface of a considerable area (~ 1 cm^2), we selected using a photo-resist. We used FP-4-04-V positive UV photo-resist spread over the fused silica substrates by spinning. The photo-resist film thickness was 800 – 1000 nm.

After the standing wave laser irradiation with the pulse energy density of ≈ 20 $mJ{\cdot}cm^{-2}$ the swellings on the photo-resist surface of ≈ 50 nm in height have been observed by AFM [figure 16 (a)]. The wet development in the KOH solution resulted in appearance of the holes in the photo-resist [figure 12 (b)] ≈ 1000 nm in depth. In addition, some number of shallow pits attributed to the action of the laser beams reflected from the back surface of the substrate and interfering with the incident ones has been observed.

After the wet development of the resist, the forming of the relief in the fused silica was performed by the following two methods: i) by wet etching in the hydrofluoric acid (HF) and ii) by Ar^+ ion etching at the dose of (4.4–5.0) $\times 10^4$ cm^{-2} and the ion energy of 2.5 – 3.0 keV for 12 hours in URM 3.279.029 ion milling machine. Figure 16 (c) presents an AFM image of the fused silica substrate surface etched through the holes in the developed photo resist film in HF. In this case, the maximum depth of the etched pits was ≈ 100 nm. In figure 16 (d) the surface of the fused silica substrate after Ar^+ ion etching is presented. Unlike etching in HF, in the case of ion etching the pit depth is greater than 200 nm. It was found that when radiation with the wavelength λ = 308 nm was diffracted by the ion-etched relief, ≈ 50% of the incident light energy was concentrated in the zero order of diffraction. Along with the estimates given in the previous section, these results indicate that Four Beam Standing Wave Pulsed Laser Lithography is a promising technology for production of the efficient two-dimensional phase masks on the fused silica substrate.

In conclusion, a periodic relief on the fused silica surface has been formed with an UV photo-resist exposed to four beam standing wave XeCl-pulsed laser radiation followed by the wet and Ar^+ ion etching. It should be noted that this technology could be applied not only to formation of the periodic gratings for integrated optics applications, but also to formation of the regular arrays of various semiconductors nanostructures, including the QD arrays defined by the lithography [74,75]. For example, the regular arrays of the GeSi nanoislands on the Si substrates were formed from an epitaxial Si/GeSi/Si(001) quantum well (QW) heterostructure by Standing Wave Laser Lithography followed by dry etching [76].

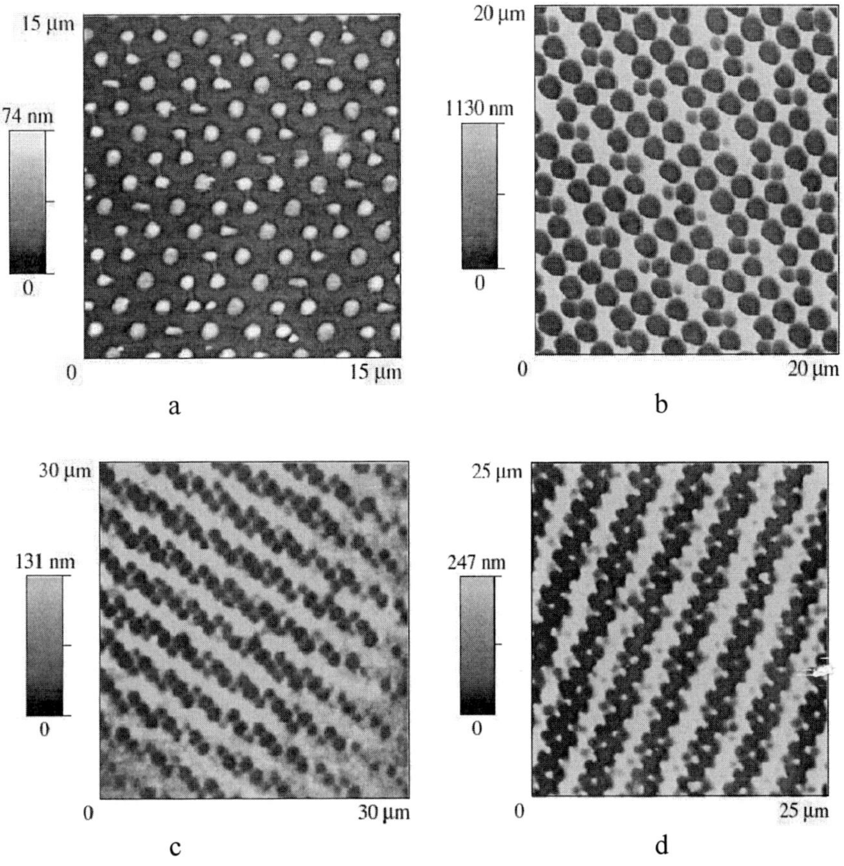

Figure 16. The AFM images of a UV photo-resist film on a fused silica substrate exposed to a standing wave pulsed XeCl laser radiation with the energy density ≈ 20 mJ·cm^{-2} (a) and after the development (b); the fused silica substrate after etching in hydrofluoric acid (c) and after Ar^{+} ion etching (d). Reprinted from [15] under kind permission by *Pleyades Publishing Co.*

Finally, some important features of Standing Wave Lithography in the case of using a photo-resist should be outlined briefly. First, let us consider the effect of the nonlinear processes in the resist during exposure, development, and processing on the resulting morphology of the processed surface and hence on the critical dimensions achievable by Standing Wave Lithography [2]. The state of art resists used in the microelectronic production are usually engineered to be highly nonlinear in order to obtain the sharp thresholds between the exposed areas and the unexposed ones. This is a really important characteristic, which has allowed the development of many of the resolution enhancement techniques, such as optical proximity correction or the phase shift masks, which are employed by the microelectronic production routinely. As a result, a sinusoidal aerial image pattern typical for Standing Wave Lithography is transformed into a square wave developed resist pattern. Let us consider a one-dimensional example. Let us assume for clarity that the standing wave intensity is given, in simple terms, by the $1 + \cos(2kx\sin\vartheta)$ function, where $k = 2\pi n/\lambda$ is the optical wave-vector in the medium above the resist and ϑ is the angle of incidence in the same medium. After development of the photo resist, a square wave line-shape is typically achieved. Taking the Fourier transformation of the latter allows us to present it in the following form:

$$z(x) = \sum_{l} \frac{\sin(2lka\sin\vartheta)}{2lka\sin\vartheta}\cos(2lkx\sin\vartheta). \qquad (27)$$

where a is the critical dimension (the width of the line) of the pattern. Expression (27) contains the spatial frequencies well above those in the aerial image clearly. It is well known that the optical system band-pass limits the pattern density in a single exposure to ~ λ/NA, where NA is the numerical aperture of the system [2]. However, the nonlinearities in the resist processing result in higher harmonics extending in spatial frequency well beyond the diffraction limits [77]. This nonlinear concept could likely to extend the reach of optical lithography by a factor of at least two. Further similar extensions by integral factors 2, 3, 4, ..., are possible. The key issue is to apply the nonlinearities between the exposures and to align the images in such as way as to reduce or eliminate the lower frequency components, leaving the higher density of the structures. This is known widely as the spatial period division or as the spatial frequency multiplication depending on whether the focus is on real space (division) or on the reciprocal space (spatial frequency multiplication). In principle, this process can be repeated multiple times, for integer spatial frequency multiplications — 3, 4, and more times. Referring to all the above statement, S. R. J. Brueck had claimed that "there is no fundamental limit to optical lithography" [78]. However, further decreasing of the critical dimensions achievable by optical lithography will meet increasing difficulties resulting from the alignment, process, uniformity, and edge roughness problems. The alignment requirements the image uniformity requirements, and the difficulties of minimizing the edge roughness of the resulting patterns will ultimately provide the practical limits for the optical lithography [79]. The optical aberrations and other contributions to the wave-front distortion will introduce positional errors and limit the ability to interpolate the patterns. These issues need to be investigated before the ultimate limitations of these spatial frequency multiplication schemes can be estimated. The alignment of the patterns produced by the multiple exposures is another important issue. For the periodic patterns provided by Standing Wave Lithography, the image being printed is a large grating, as is well known, the angular resolution of a grating scales as the number of lines in the grating, so that the alignment schemes based on the moiré interference offer a promising approach to the alignment between the steps of the spatial frequency multiplication based process [80].

Using an immersion medium represents another way to further decreasing of the critical dimensions achievable by Standing Wave Lithography. A simple addition of a prism to the inter-ferometric setup shown in figure 2 (b) allows introduction of an immersion medium between the prism and the resist film. Water is a quite suitable immersion fluid for the UV wavelength band. The refractive index of water is 1.44 at at the wavelength of 193 nm [81].In the simplest configuration, a droplet of distilled water cam be placed between the processed wafer surface and an uncoated prism thus forming an almost perfect optical contact [2]. It is worth noting that such setup could be used in the direct standing wave modification as well in many cases.

In the recent years, the semiconductor device industry has been following a 15-year plan, referred to as International Technology Roadmap for Semiconductors (ITRS) [82], which predicts scaling of the critical dimensions of the semiconductor technology according to Moore's law (0.7 linear dimension shrink every three years). For the periodic patterns, which are used widely in many Si-based IC manufacturing processes, the 45-nm half-pitch, which is planned to introduce in 2010, can be achieved using Standing Wave Lithography with a 193-nm excimer laser source. For the 32-nm critical dimension, which is planned to achieve in 2013, frequency doubling with the H_2O immersion result is suitable. Using the frequency tripling and water immersion, the critical dimension is as low as 16 nm, which is planned to be introduced in 2019. Further development could probably be provided using an immersion medium with higher refractive index in the relevant wavelength band such as HI (and a higher index final glass element as well) [2].

At the same time, using a resist and/or an immersion medium results in some problems related to the light propagation in the above mentioned medium. The process of the light propagation in the resist or in the immersion fluid depends on its polarization strongly, especially at the steep incidence angles [83]. For the TE polarization, as follows from Eq. (1), the pattern contrast $(I_{max} - I_{min})/(I_{max} + I_{min})$ equals to unity regardless to the pattern period. In contrary, for the TM polarization, the image contrast decreases with decreasing pattern period as a result of the competition between the out-of-phase E_x^2 and E_z^2 components to the total intensity. At the interference angle of 45° in the resist or in the immersion medium, these two components are equal to each other in the magnitude and are opposite to each other in the phase. Consequently, the pattern contrast tends to zero in this case. For larger interference angles, the line-space pattern becomes inverted. So far, the reduction of the interference pattern period results in the loss in the image contrast at the same time. This means that in order to take the advantages of the resist nonlinearity and/or immersion to reduce the pattern period while keeping the image contrast at the same time, a control of the interfering beams' polarizations is necessary.

Chapter 4

The Two-Dimensional Periodic Patterning of the Solid Surfaces by Direct Four-Beam Standing Wave Excimer Laser Modification

In this chapter, the results of the investigations of direct four-beam standing wave-pulsed excimer laser patterning of the metal films on Si substrates, of the synthetic polycrystalline diamond films on the single crystal diamond substrates, and of the polyimide and titanium dioxide films on the fused silica substrates are presented.

A. The Fe—C Films on Si Substrates

In this section, we present the results of the periodic patterning of the Fe—C films with the thickness of several tens of nanometers on the Si substrates by direct Four-Beam Standing Wave-Pulsed Laser Ablation. These studies continue the series of investigations aimed at the development of the technology for the fabrication of the regular arrays of the ferromagnetic nanoparticles with the submicron period by modifications of the thin metal films by direct four beam standing wave XeCl excimer pulsed laser modification [84,85]. In this section, the forming of the two-dimensional periodic arrays of the holes in the Fe—C films with different shape and periods in different directions is reported. The effect of the geometry of the optical scheme of the inter-feromentric laser setup on the nanostructure morphology has been studied. It was demonstrated that the two-dimensional periodic arrays with the variable geometrical parameters of the elements can be formed by varying the orientation of the bisectors of the angles between the laser beams with respect to the modified sample surface. The possibility of varying the incident angles and, so far, the period(s) of the arrays and/or the ellipticity ratio of the formed nanoelements is an important advantage of the four-beam standing wave lithographic technique useful for some applications. An example of such applications is the periodic arrays of elongated ferromagnetic nanoparticles for dense magnetic data storage elements [6]. The features of the nanostructure morphology were shown to be highly sensitive to the duration and energy of the laser pulses. The conditions for the forming of near perfectly periodic two-dimensional arrays of the submicron structures in the thin Fe—C films had been determined.

The main processes responsible for the modification of the metal film surface are usually determined by the heat release process. As it was demonstrated in Chapter II, the submicron spatial modulation of the temperature with a considerable modulation depth can be achieved when the duration of the laser pulses does not exceed several nanoseconds. This requirement dictates the choice of the operation mode of the excimer laser used for the metal film patterning.

In figure 17 an AFM image of the surface of a Fe—C film on a Si substrate modified by a single laser pulse is presented. Figure 17 (c) presents a surface profile across a pit along the line shown in figure 17 (b). The width and the depth of the pits were ≈ 0.8 μm and ≈ 25 nm, respectively. The ellipticity of the pits was ≈ 0:3 that is close to the ratio of the interference pattern periods in the respective directions. The ratio of the size of the modified area to the period of the array for typical experimental conditions was ≈ 0:25.

An important feature of the pit morphology is a swell around each pit. The AFM measurements revealed that approximately ¼ of the material removed in the process of the pits' formation was transferred to the swells. The rest material was ablated.

The morphology of the swells has been found to depend on the laser beam parameters strongly. In figure 18 the AFM images and the surface profiles of the pits formed at different laser pulse energy densities are presented. The samples shown in figure 18 (a) and (c) were formed at the laser pulse energy densities in the maxima of the interference pattern of ≈ 0.3 $J{\cdot}cm^{-2}$ and ≈ 0.5 $J{\cdot}cm^{-2}$, respectively. Both the depth and the width of the pits increased with increasing pulse energy density. In the samples shown in figure 18 about 20%

of the material removed from the film was transferred into the swells around the pits.

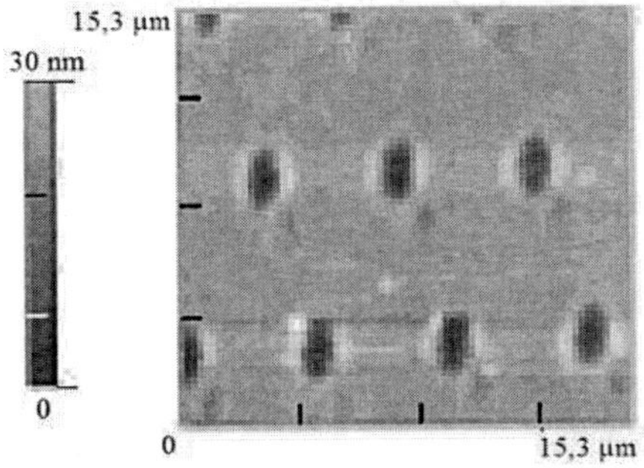

(a)

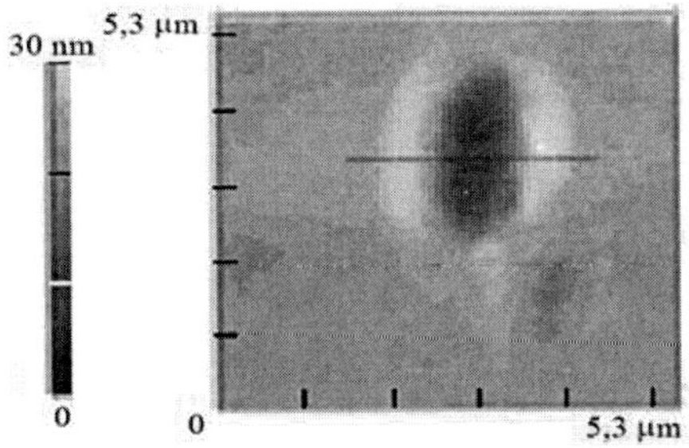

(b)

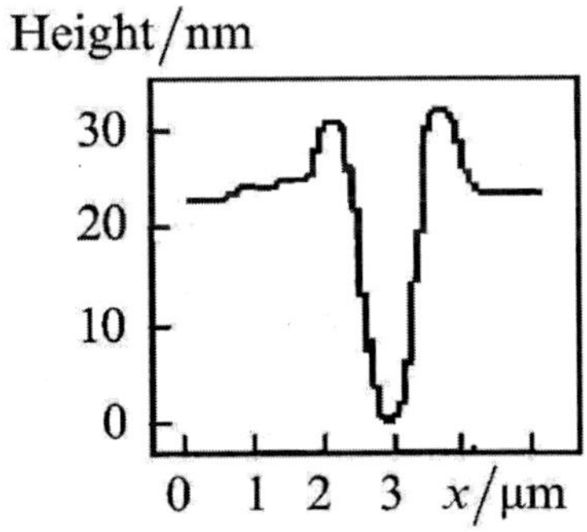

(c)

Figure 17. The AFM images (a, b) and the surface profile (c) of a Fe—C film on a Si substrate patterned by Standing Wave Pulsed Laser Ablation. Reprinted from [6] under kind permission by *Turpion Co. Ltd.*

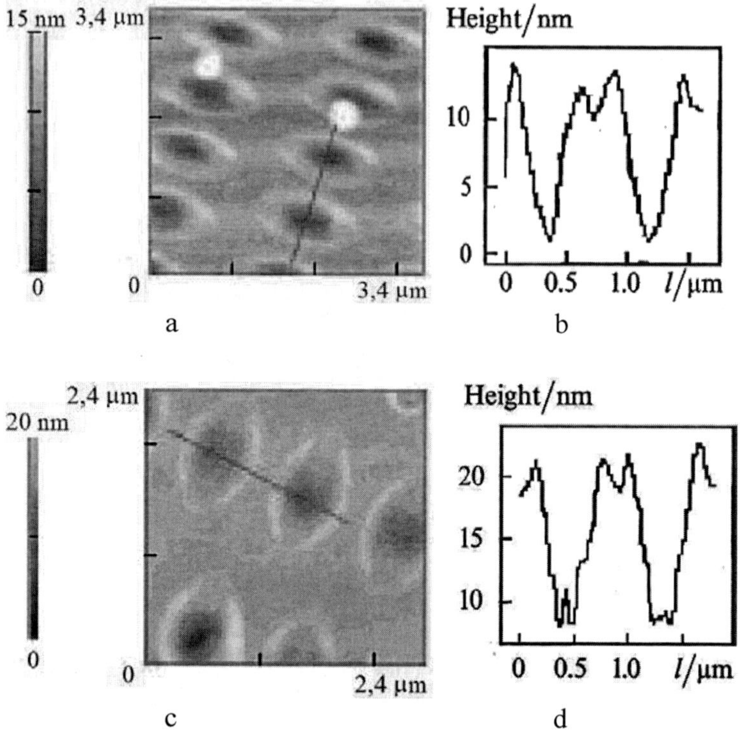

Figure 18. The effect of the laser pulse energy density on the morphology of the pits in the Fe—C film on the Si substrate patterned by Standing Wave Pulsed Laser Ablation. The laser pulse energy density, J cm^{-2}: (a, b) — 0.3; (c, d) — 0.5. Reprinted from [6] under kind permission by *Turpion Co. Ltd.*

The laser pulse duration in the experiments on the modification of the samples shown in figure 18 was selected to be short enough (≈ 3 ns) to ensure a weak heat dissipation during the laser pulse. In figure 19 an AFM image and the surface profiles of a Fe—C film modified at the laser pulse duration of ≈ 7 ns and the pulse energy density of ≈ 0.44 J·cm^{-2} are presented. Comparing the AFM image in figure 19 to the ones presented in figure 18 one can see that the width and the height of the swells around the pits are much greater in the former case so that the swells around any two adjacent pits even tend to overlap [see figure 19 (b)]. The ellipticity of the pits in figure 19 was ≈ 0.5 although the ratio of the periods in the interference pattern of the laser radiation was ≈ 0,25.

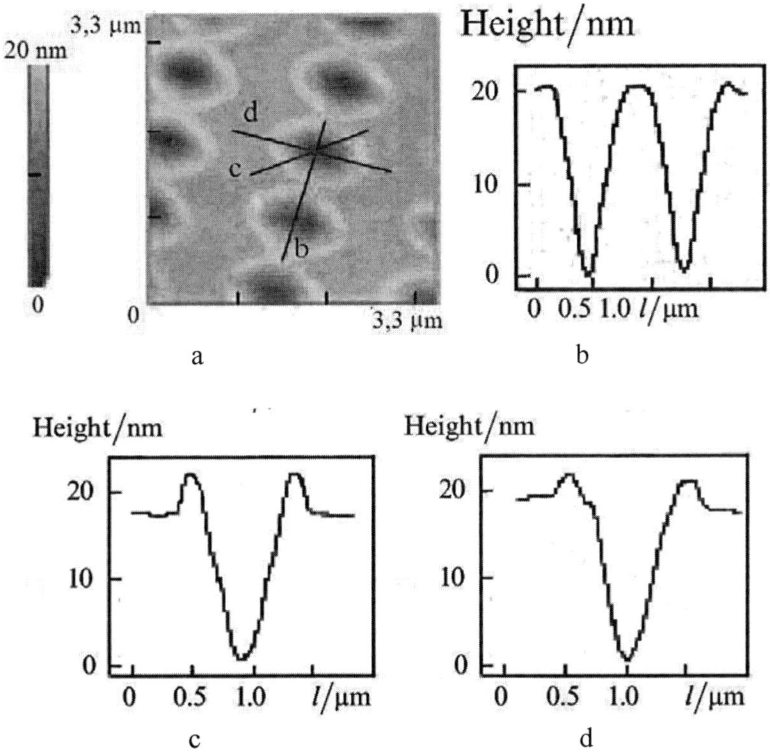

Figure 19. An AFM image (a) and the surface profiles (b — d) along the lines labeled with the respective letters in the AFM image (a) of a Fe—C film on Si substrate patterned by Standing Wave Pulsed Laser Ablation under conditions of sufficient heat dissipation during the laser pulse. Reprinted from [6] under kind permission by *Turpion Co. Ltd.*

The results presented in figure 18 and figure 19 allowed us to estimate the heat diffusion length to be ≈ 0.25 μm that is consistent with the results of the calculations presented in Chapter II. The maximum temperature of the film in the interference maxima can be estimated to be close to the boiling point of iron. The parameter of the two-phase destruction [52] $\chi = (h/D)(v_l/v_v) < 0.7$ where h is the film thickness, D is the diameter of the heated area, v_l is the velocity of melt flowing out of the irradiated area, and v_v is the evaporation rate. The calculated ratio of the mass of the removed liquid metal to the mass of the removed vapor is less than 0.3. The latter estimate agrees well with the measured ratio of the volume of the swells around the pits to the volume of the entire pit [figure 17 (c)]. Knowing the mean velocity of the melt rolling [52] ($v_l \approx 250$ m s^{-1}) and the pit radius ($r \approx 0.5$ μm) one can estimate the time of rolling of the liquid metal to be ≈ 2 ns that is close to the duration of the laser pulse. Note that the role of the heat diffusion under the conditions of the experiments described here becomes much less important when both periods of the interference pattern exceed ≈ 2 μm (see figure 17).

Thus, based on the study reported, we have determined the conditions necessary for the formation of the nearly perfect two-dimensional periodic submicron arrays of the pits in the Fe—C films on the Si substrates by Four Beam Standing Wave-Pulsed Laser Ablation. The primary mechanism of the pits' formation was associated with the two-phase destruction of the metal film material. In figure 18 the AFM images of the Fe—C films on the Si substrates patterned by Standing Wave-Pulsed Laser Ablation with different values of the ratio of the periods of the interference pattern in different directions Q_i are presented. It is worth noting a remarkable uniformity of the patterned film surface over the whole AFM images.

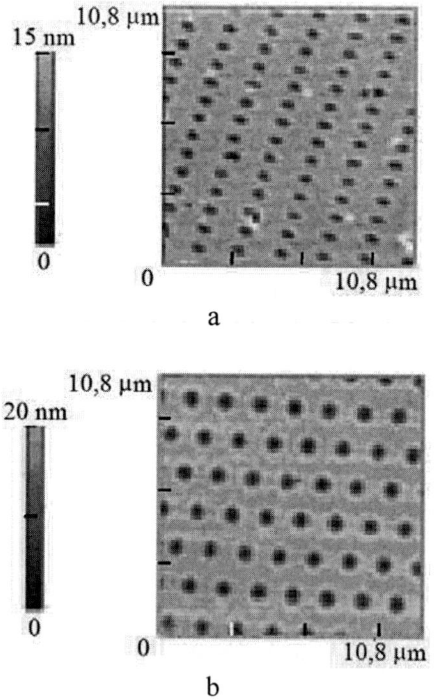

Figure 20. The AFM images of the Fe—C films on Si patterned by Standing Wave Pulsed Laser Ablation. The ratio of the periods of the interference pattern in different directions Q_i: *a* — 0.25; *b* — 0.5. Reprinted from [6] under kind permission by *Turpion Co. Ltd.*

One of the promising applications of the metal films patterned by Standing Wave-Pulsed Laser Ablation is the nanoscale magnetic structures. There is a substantial interest in the magnetic properties of the individual single-domain nanomagnetic structures originating from originating from their potential for application in the magnetic data storage with increased density, in spintronics, etc. [86,87,88].

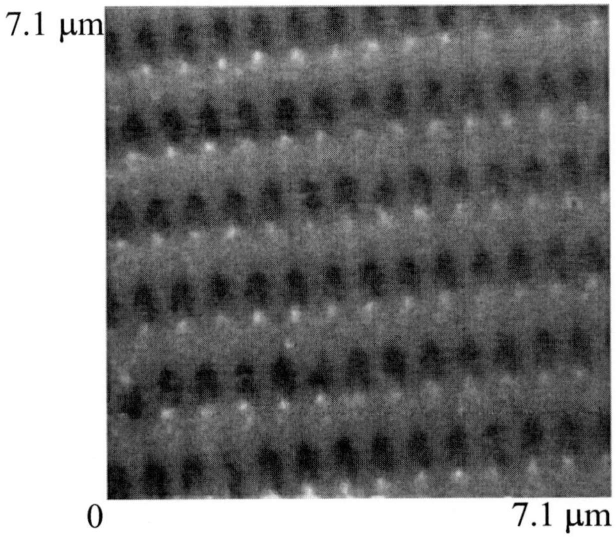

Figure 21. A MFM image of the surface of a $Fe_{0.7}Cr_{0.3}$ filam modified by Four Beam Standing Wave Laser Ablation after the magnetization along the longer axes of the ellipsoidal-shaped holes. The energy density was 250 mJ/cm^2. The image is obtained for a residual state. Reprinted from [57] under kind permission by *Pleyades Publishing Co.*

Standing Wave Lithography is a natural technology for the fabrication of the large area arrays of the single domain magnetic nanoparticles [89,90,91,92]. The interest in this subject has been stimulated also by the development of the lithographically defined multilayered elements such as the magnetic tunnel junctions [93] or the spin-valve structures [94,95], which are the key components in the state-of-art high-density magnetic memory.

In figure 21 a Magnetic Force Microscopy (MFM) image of the surface of a $Fe_{0.7}Cr_{0.3}$ film modified by Four Beam Standing Wave Laser Ablation at the energy density of 250 mJ/cm^2 is presented [57]. The image has been measured after an in-plane magnetization along the longer axes of the ellipsoidal shaped holes. The asymmetrically-shaped magnetic dipoles directed along the larger hole's axes are seen clearly.

The perforated superconducting films also demonstrate some interesting phenomena. The holes act as the pinning centers for the quantized magnetic flux [2]. At low magnetic fields, the flux lines are all associated with these pinning holes and the magnetization shows distinctive structure when the flux density is commensurate with the density of the hole array or is an integral multiple of the hole density so that the flux lattice forms a superstructure locked to the hole array [96]. For a large-area sample, mismatches of the crystal structure between grains nucleated in different regions of the sample like the grain boundaries in crystal growth have been observed [97].

The optical materials with a negative refractive index (so-called metamaterials) [98] is another promising application of the metal films with the regular alrrays of holes made by Four-Beam Standing Wave Pulsed Laser Ablation. The interest in this topic has increased dramatically as a result of the prediction of important new effects, including diffractionless imaging, in such a material [99]. The mechanism of Negative Refractive Iindes is as follows [2]. The metals provide a negative permittivity at frequencies below the plasma resonance as a result of the free electron response to the incident electromagnetic field. On the other hand, there is no comparable natural negative magnetic susceptibility in the metals. However, the *LC* resonant circuits provide a band of negative permeability around their resonance frequency. In this scope, an array of such resonant microstructures with the sizes much less than the wavelength of the electromagnetic wave distributed inside a dielectric matrix can act effectively as a medium with a negative permeability over a certain frequency range. In this case, the permeability of such a material depends strongly upon the properties of the individual *LC* resonant circuits but is independent upon their arrangement, e.g. on the periodicity. Such materials, permittivity and permeability of which results from the artificially fabricated resonant microstructures, are classified as the metamaterials.

An example of such a structure fabricated using Standing Wave Lithography was described in [100]. The structure consisted of an $Au/Al_2O_3/Au$ film stack on a glass substrate; the Au films were perforated by Standing Wave Lithography in such a way that the holes in the Au films were aligned vertically (i.e., in the direction normal to the substrate). The hole array pitch was ≈ 0.8 μm. This stack demonstrated a negative refractive index in the near IR band at the emission wavelength of ≈ 2 μm. The magnetic response of such structures takes place at certain polarization of the incident light as a result of the distributed inductance/capacitance of the two perforated metal films separated by a dielectric layer. Later, the optimized structures have been designed to provide an improved negative index material [101].

Standing Wave Lithography is especially promising for the fabrication just of this particular class of the negative refractive index structures. There is also a considerable number of works reported in the literature where the resonant microstructures (so called nanoantennas) were fabricated by E-Beam Lithography [102]. Therefore, the size of the investigated samples has been restricted. In the case of Standing Wave Lithography, the size of the samples is limited by the convenience of handling rather than by the cost and the time of the metal film patterning. The example described is a good illustration of the major advantages of Standing Wave Lithography in fabrication of the nanophotonic structures.

Direct Four-Beam Standing Wave-Pulsed Laser Ablation is ideally suited to the fabrication of the large two-dimensional arrays of the holes in the metal films, as has been amply demonstrated in this section. The fabrication process in this case is simply reduced to deposition of a metal film onto a substrate, a single shot four-beam laser patterning of the film, deposition of a dielectric layer onto the patterned metal film, and, finally, repetition of the metal film deposition and patterning. All these stages could be easily combined in a single cycle e.g. in a vacuum chamber for Magnetron Sputtering or for Electron Beam Evaporation. This simplicity of the fabrication of the large areas as well as a wide choice of the materials is highly favorable both for the extensive studies of the negative refractive index effects in such structures and also for fabrication of the large area nanoscale structures for device applications.

In addition, t is worth noting that the arrays of the holes in the metal films obtained by Direct Standing Wave Laser Ablation can be transformed in the arrays of the ring-shaped nanostructues just by ion sputtering of the metal film remaining between the swells around the holes leaving the swells not sputtered completely.

B. The Polyimide Films on the Fused Silica Substrates

The objects to be modified were 1-μm-thick polyimide films deposited onto the fused silica substrates.

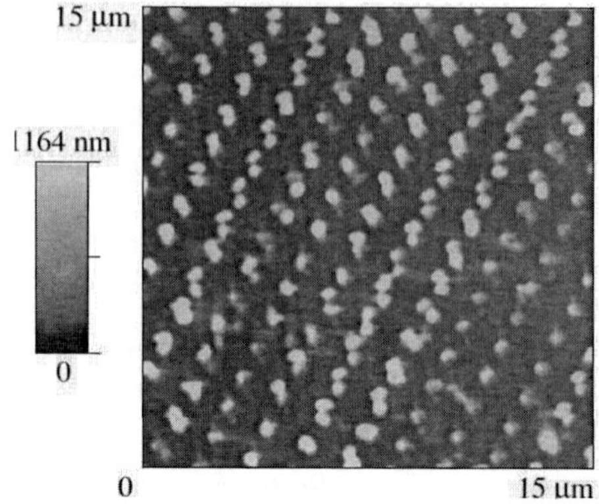

a

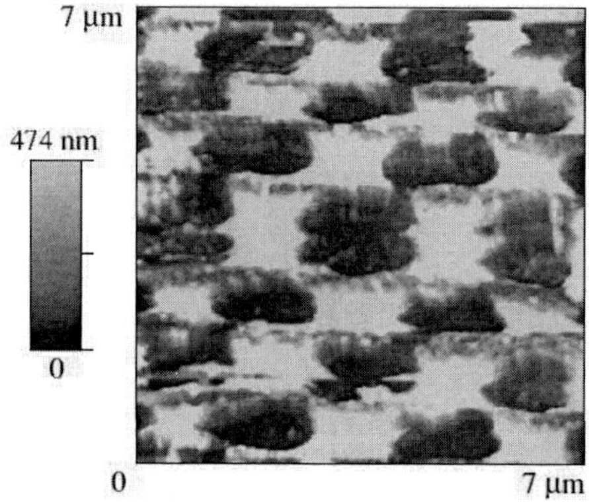

b

Figure 22. The AFM images of the polyimide films on the fused silica substrates modified by the four beam standing wave pulsed laser radiation at the pulse energy densuty of (a) 60 mJ cm^{-2} and (b) 120 mJ cm^{-2}. Reprinted from [15] under kind permission by *Pleyades Publishing Co.*

The experiments on the standing wave-pulsed laser modification of the polyimide films have shown that there exist two different modes of the modification subject to the pulse energy density. At the pulse energy densities less than 100 mJ cm^{-2}, the swelling of the film surface has been observed, the most probably originating from the destruction of polyimide. In figure 22 (a) an AFM image of the polyimide film on the fused silica substrate modified at the pulse energy density of $\approx$ 100 mJ cm^{-2} is presented. The height of the swellings depended on the pulse energy density and ranged from 300 to 1000 nm. At the energy densities greater than 100 mJ·cm^{-2}, the regular arrays of the holes in the polyimide film have been observed. In figure 22 (b) an AFM image of the polyimide film modified in the regime specified above is presented. Such objects can be used as the resist masks in the submicron lithography. Unlike the conventional Standing Wave Lithography where such masks are formed in the two-step process (the first step is the exposing of the resist and the second one is the development), direct Standing Wave-Pulsed Laser Ablation allows a single-step mask formation that is highly appreciable from the viewpoint of the cost reduction.

C. Polycrystalline Diamond

Diamond has been being considered to be a promising wide-gap semiconductor material for decades. In particular, the ultrathin epitaxial films of the synthetic diamond are considered now to be a promising material for some advanced microelectronic applications. Diamond is featured by outstanding thermal and radiation hardness as well as by high thermal conductivity. The latter property is extremely valuable for the very large scale integrated circuits where the problem of the heat sink is now a critical one.

Here it was demonstrated that a regular periodic relief can be formed on the surface of the synthetic diamonds by direct Standing Wave Pulsed Laser Ablation. The importance of this result follows from the fact that radiation and chemical hardness of diamond causes considerable problems in its processing by the standard microelectronic techniques (such as dry etching, etc.).

In figure 23 the AFM images of the surface of the synthetic polycrystalline diamond exposed to the standing wave pulsed laser radiation with the energy density of ~ 10 $J \cdot cm^{-2}$, which corresponds to the radiation induced damage threshold for the single crystal diamond are presented. At such an energy density, different places of the policrystalline diamond surface exhibited different types of damage. There were the areas where the damage pattern correlated with the standing wave one well [figure 23 (a)]. These were the surfaces of single crystal grains in the polycrystalline diamond film. In some places, a severe damage independent on the incident radiation spatial structure has been observed [figure 23 (b)]. These places were associated with the inclusions of graphite between the diamond grains highly absorbing the laser radiation while the single crystal diamond is transparent to the 308-nm radiation.

In figure 23 (c) a fragment of the modified polycrystalline diamond surface between the single crystal grains where the absorbing graphite layer and, so far, the severe damage were absent is presented.

In conclusion, it was demonstrated that using the direct Four Beam Standing Wave-Pulsed UV Laser Ablation, one can form a two-dimensional periodic relief with a submicron period on the diamond surface. In particular, this technology could be applied for secret marking of the diamonds used in jewelry.

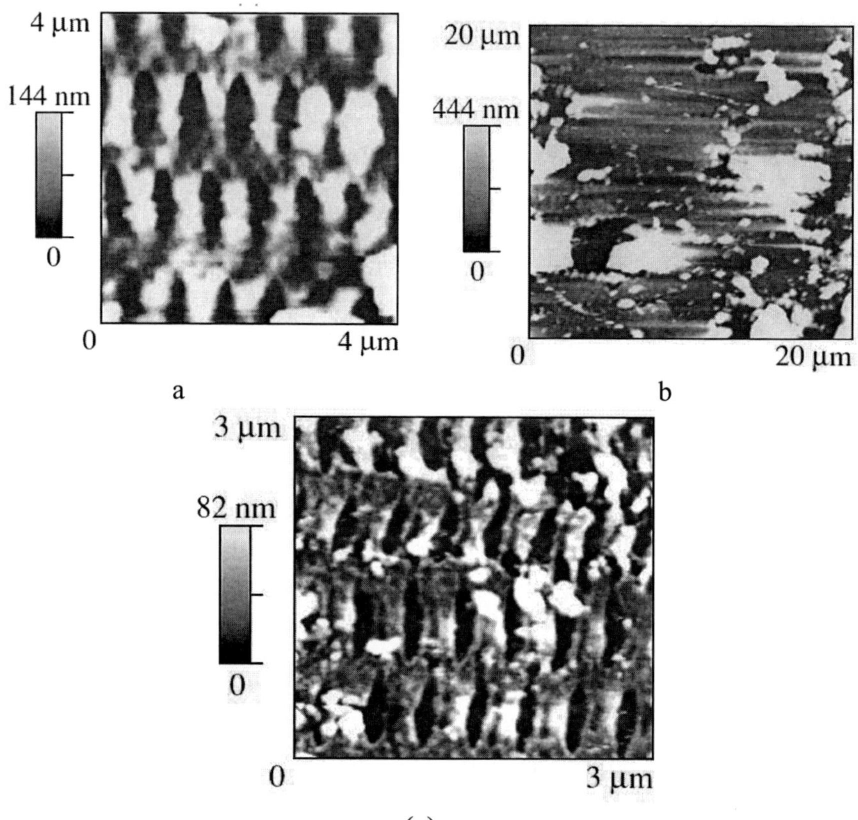

Figure 23. The AFM images of the modified surface of synthetic polycrystalline diamond: (a) the modified surface of a single crystalline grain; (b) severe damaged surface of a graphite inclusion between the crystalline grains, and (c) at the boundary between two adjacent single crystal grains. The pulse energy density was ≈ 10 mJ/cm^2. Reprinted from [15] under kind permission by *Pleyades Publishing Co.*

D. Titanium Dioxide

The properties of titanium dioxide have been being studied extensively for several last decades since this material is very promising for various applications. Among them, the development of highly efficient and cost effective solar cells should be mentioned [103,104,105,106,107,108].

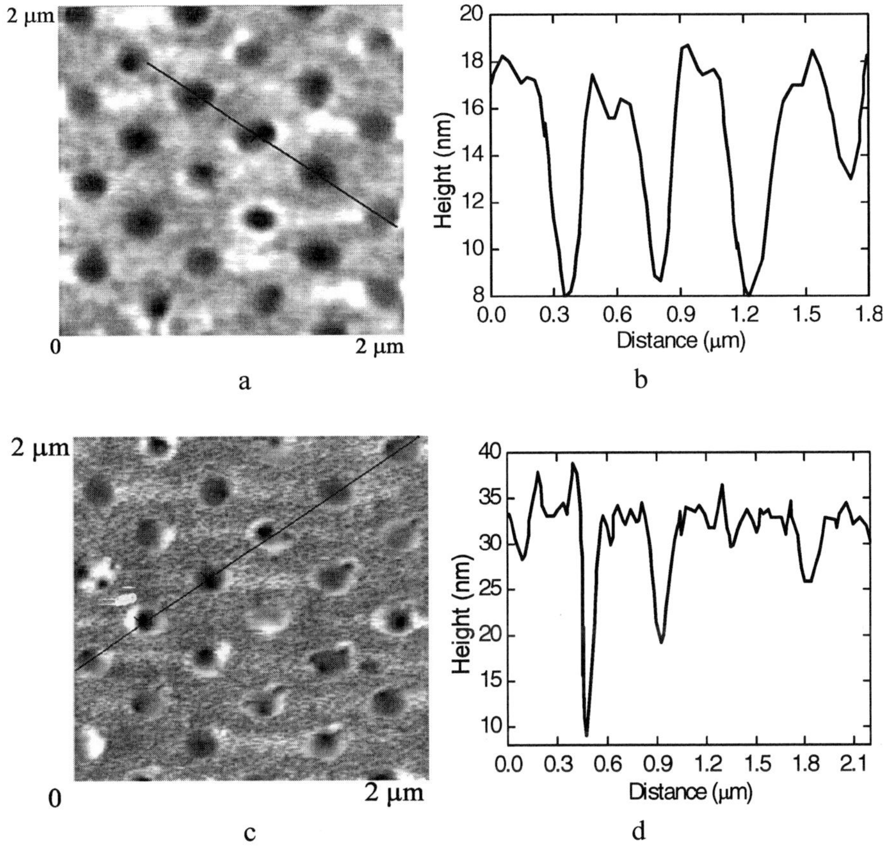

Figure 24. The AFM images (a, c) and the surface profiles (b, d) of a TiO_2 film on a fused silica substrate patterned by a single pulse of the four-beam standing wave UV laser radiation.

Another promising direction is biocatalysis [109,110,111,112,113], including the hydrogen generators employing water decomposition on TiO_2 [114,115,116,117]. Titanium dioxide is featured by high chemical durability, by high tolerance to the human body tissues, and by high refractive index (> 2) in the visible and IR wavelength bands. All these properties make titanium dioxide promising for the molecular separators, the diffraction gratings, the photonic crystals, etc. Many of the applications mentioned above require porous TiO_2 films, which are produced using the ion beams or the solid masks. It is well known that the developing of the surface of the catalysts improves their performance.

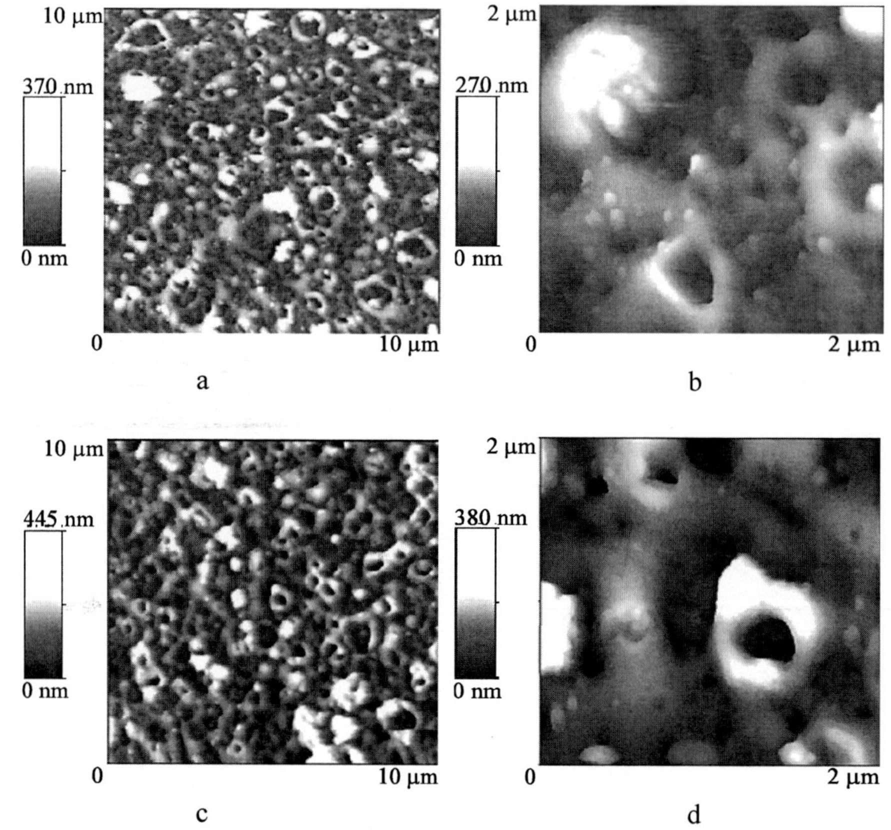

Figure 25. The AFM images of a TiO_2 film on a fused silica substrate patterned by a series of two pulses of the four-beam standing wave UV laser radiation.

In the investigations presented in this section, we have employed the direct four-beam standing wave XeCl excimer laser modification for formation of the nanoporous TiO_2 films. The initial TiO_2 films to be patterned have been formed on the fused silica substrates by sol-gel technology. The film thickness ranged from 300 to 500 nm, the annealing temperature was 250 ÷ 400°C. The following two regimes of the nanoporous TiO_2 film formation has been explored. In the first case, the TiO_2 films were patterned by a single laser pulse. In the second mode, the TiO_2 films were subjected to a series of two laser pulses.

In figure 24 the AFM images and the surface profiles of a TiO_2 film subjected to a single laser pulse is shown. The depth of the pits on the patterned film surface was less than 30 nm. In this case, the increase in the film surface area is insufficient.

Another pattern of the surface morphology has been observed when the TiO_2 film was patterned by two laser pulses (figure 25). In this case, the height differences on the film surface were more than 300 nm that resulted in the increase of the surface area by the factor of ≈ 1.5.

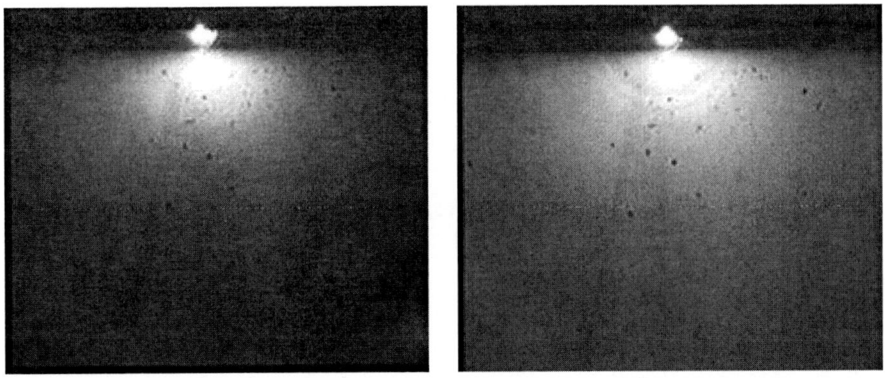

a b

Figure 26. The electron diffraction patterns from a TiO_2 film on a fused silica substrate before (a) and after (b) four-beam standing wave UV-pulsed laser modification.

The initial TiO_2 film had amorphous crystalline structure. After the four beam laser patterning, the crystallization of the TiO_2 films has been observed. In figure 26 (a) and (b), the electron diffraction patterns of a TiO_2 film before and after the laser patterning, respectively are presented. Before the laser patterning, a halo peculiar to the amorphous materials has been observed only [figure 26 (a)]. After the laser treatment a diffraction pattern peculiar to the polycrystalline materials has been observed [figure 26 (b)]. Crystallization of the amorphous TiO_2 films during the four-beam-pulsed laser patterning can be applied to improve their photo catalytic properties.

Chapter 5

SELF-ASSEMBLING OF THE PERIODIC ARRAYS OF THE NANOMETER-SIZED SEMICONDUCTOR ISLANDS STIMULATED BY STANDING WAVE UV LASER RADIATION

In this chapter, we present the results on the formation of the two-dimensional regular periodic arrays of the semiconductor surface nanoislands formed by self-assembling stimulated by four coherent interfering pulsed laser beams incident onto the silicon and gallium arsenide single crystal wafers. The lateral size of the islands ranged from 60 to 100 nm and the height from 6 to 12 nm that was 5 – 10 times smaller than the period(s) of the standing wave. The island formation was attributed to the effect of the elastic strain at the crystal surface induced by the non-uniform heating of the substrate by the standing wave laser radiation. The effect of the laser beam parameters including the polarization of the interfering laser beams on the resulting nanoisland array morphology has been investigated. It has been already mentioned above (see Chapter II) that the distribution of radiation in the standing wave field depends on the polarization of interfering laser beams significantly. The analytical expressions and the computer calculations of the spatial distribution of the standing wave intensities for various polarization configurations can be found in [43]. In our experiments, both the TE—TE configuration and the TE—TM one had been explored. The conditions for obtaining the uniform nearly perfectly periodic arrays of Si and GaAs nanoislands on the Si(001) and GaAs(001) substrates, respectively, had been determined.

A. Silicon

The results presented in this subsection were obtained using single laser pulses with the duration of 10 ns and the energy density of (1.0 – 1.5) J·cm^{-2}. In figure 27, a typical AFM image of Si(001) surface patterned by four-beam standing wave UV pulsed laser radiation in the TE—TE polarization configuration is presented. The standing wave pattern in this case corresponds to the two series of stripes perpendicular to each other. The AFM image in figure 27 (a), as well as the two surface profiles along the lines *1* and *2* in the main image [figure 27 (b) and (c), respectively], demonstrate the nanoislands with the diameter of 40 – 60 nm to form in the maxima of the standing wave intensity. The islands were separated by the almost flat regions of the substrate with the dimensions of ≈ 30 × 150 nm^2. An essential feature of the self-assembling in this particular regime is the appearance of the pits of various shapes and dimensions near the islands. The pits have lateral dimensions more than 30 nm and the depth of 4 – 6 nm. As can be seen from figure 27, the pits were located at different sides of the neighboring islands.

This pattern of modification of the crystal surface differs significantly from the results of the standing wave laser modification of the amorphous materials (see Chapter IV, Sec. A) where the symmetric swells were formed around the pits [6]. A substantially different character of self-assembling was observed in the case of using the beams with the TE—TM polarization configuration. In this case, the sites of maximum intensity are well localized and surrounded by the lines of zero intensity. As a result of such an intensity distribution, the self assembled Si nanoislands are defined better in this case.

In figure 28, a typical AFM image and a surface profile of Si(001) surface patterned in the TE—TM polarization configuration is presented. The AFM image revealed a nearly perfect two-dimensional periodic array of islands with the diameter below 100 nm and the height of ≈ 12 nm. It should be noted that all islands in the array are located at the same side of the adjacent pits that indicates that the conditions of diffusion for silicon atoms were the same over the whole AFM image.

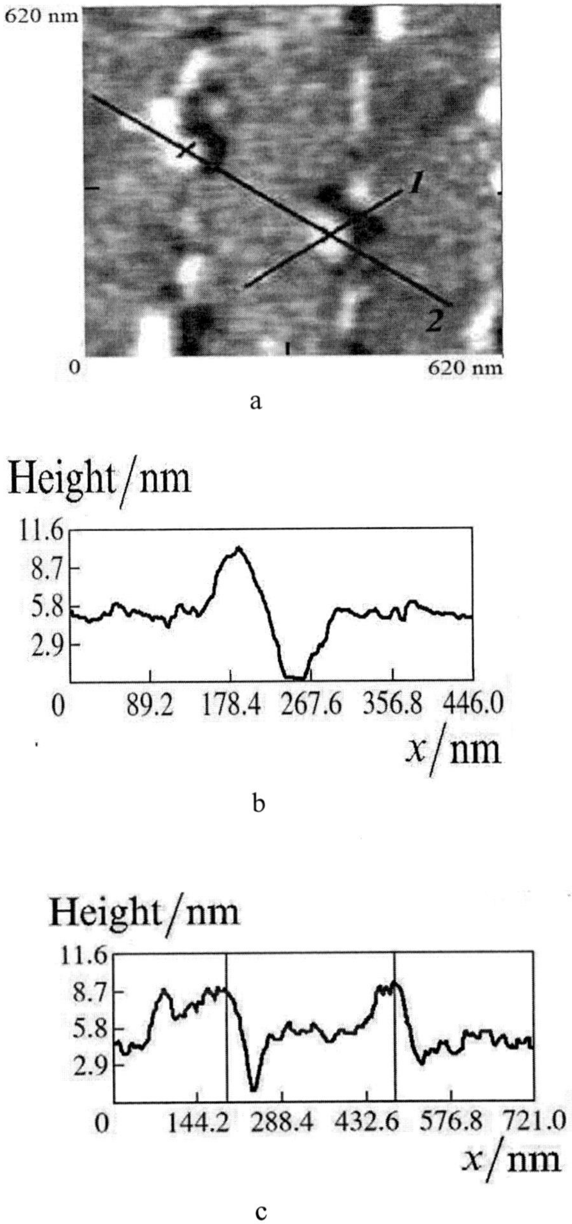

Figure 27. An AFM image (a) and the surface profiles (b, c) of Si(001) surface patterned by the four-beam standing wave UV-pulsed laser radiation in the TE–TE polarization configuration. Reprinted from [37] under kind permission by *Pleyades Publishing Co.*

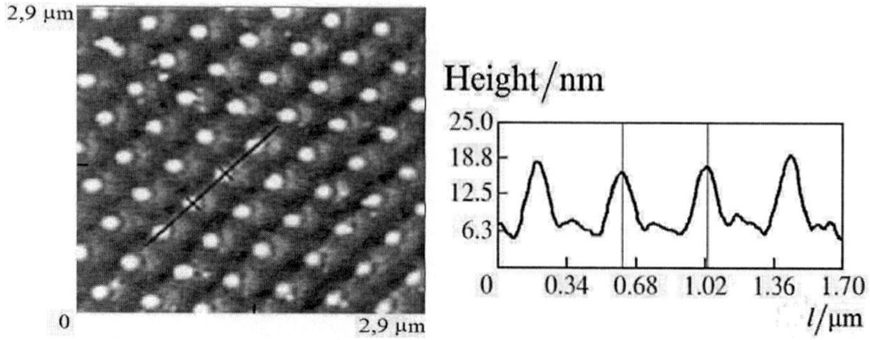

a b

Figure 28. An AFM image (a) and a surface profile (b) of Si(001) surface patterned by the four-beam standing wave UV-pulsed laser radiation in the TE—TM polarization configuration. Reprinted from [37] under kind permission by *Pleyades Publishing Co.*

B. Gallium Arsenide

In figure 29 an AFM image of GaAs(001) single crystal surface patterned by four-beam standing wave UV-pulsed laser radiation in the TE—TM polarization configuration is presented. The islands observed on the GaAs surface had the lateral dimensions of ≈ 60 nm and the height of ≈ 7 nm. The surface profile showed that the positions of the islands with respect to the interference pattern knots scattered within ≈ 30 nm.

The results presented here demonstrate that the minimum lateral size of the islands was 5 – 10 times smaller than the period of the standing wave of the laser radiation. In the case of laser modification of the amorphous or polycrystalline films [15, 118], the typical size of the region of modification usually amounts to 50% of the standing wave period. This significant difference between the results of the laser modification of the single crystal and of the amorphous surfaces leads to a conclusion that the resulting surface morphology in the laser modification of single crystals is affected by the inhomogeneous laser-induced strain field arising from the local heating of the crystal surface by the standing wave radiation. This heating results, first, in the thermal expansion of the near-surface layers inducing the elastic stress, and second, in the stimulation of the diffusion of the atoms of the modified material in the elastic strain field developed in the surface layer of the crystalline substrate. The simultaneous action of the two factors listed above makes possible the formation of the nanoislands with the sizes significantly smaller than the standing wave period. Note that the island formation cannot be explained by the effect of the thermal threshold for the laser modification only since the nanoisland size remained almost the same when the energy density of the laser radiation increased two times. So far, such a modification is possible in the presence of the surface stress only.

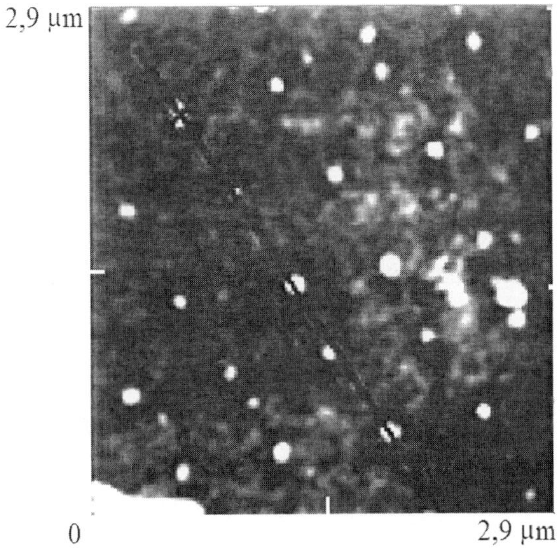

(a)

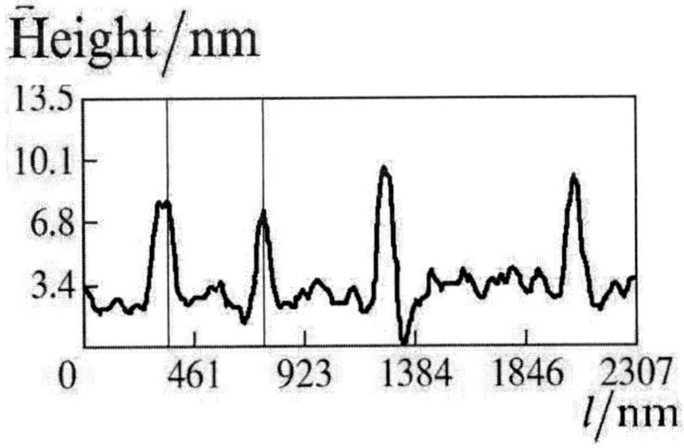

(b)

Figure 29. An AFM image (a) and a surface profile (b) of GaAs(001) surface patterned by four-beam standing wave UV-pulsed laser radiation in the TE—TM polarization configuration. Reprinted from [37] under kind permission by *Pleyades Publishing Co.*

C. InGaAs/GaAs Eptiaxial Heterostructures

In this section, the results of the four beam standing wave-pulsed laser modification of the strained GaAs/InGaAs/GaAs quantum well (QW) heterostructures are presented. These studies were the further development of the ones described in the previous sections and were aimed at confirming the suggestion of the laser-induced local thermal expansion of the surface as the key driving force for the nanoisland formation. In these purposes, the GaAs/InGaAs/GaAs hererostructures where the elastic stress was present *ab initio* were grown on GaAs(001) substrates by Molecular Beam Epitaxy (MBE).

The experiments on modification of the GaAs/InGaAs/GaAs(001) epitaxial heterostructures were carried out on the four samples. The *n*-GaAs(001) epi-ready substrates were used for the epitaxial growth by all-solid-source MBE. After the thermal removing of the native oxide from the GaAs substrate surface in the As flow, the 100 nm GaAs buffer layers were grown at 580°C followed by deposition of the 3 nm thick $In_{0.35}Ga_{0.65}As$ QWs at 520°C. Finally, the GaAs cap layers of 5, 10, 20, and 40 nm in thickness were grown at the same temperature as the $In_{0.35}Ga_{0.65}As$ layer. The conditions of the four beam laser modification were the same as in the experiments on modification of the single crystal Si and GaAs wafers (see Sec. A and B for details) but the laser pulse duration was varied within the range from 10 to 15 ns.

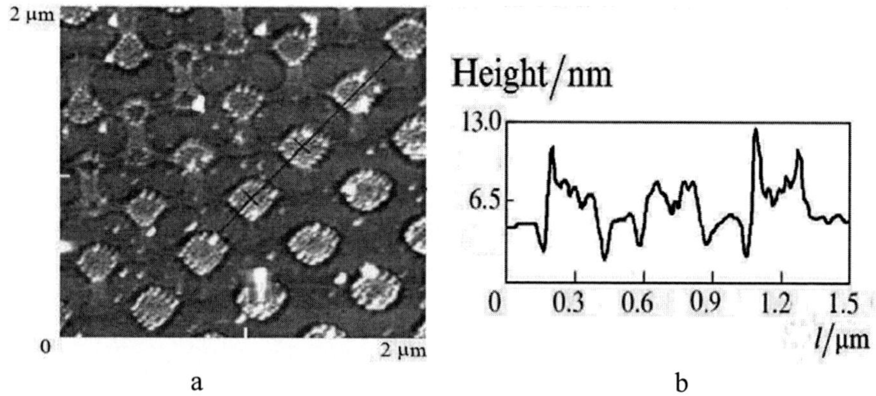

Figure 30. An AFM image (a) and a surface profile (b) of the GaAs/InGaAs/GaAs heterstructure surface modified by four beam standing wave XeCl-pulsed laser radiation.

Within the accuracy of the AFM measurements, the general pattern of the modified surface morphology did not depend on the GaAs cap layer thickness. Below the results of the laser modification of the heterostructure with the 20 nm thick cap layer are presented.

When the TE—TM polarization configuration was selected, the intensity distribution in the plane of the sample surface was treated in the first-order approximation as the sinusoidal one along both directions [43]. In figure 30 (a) and (b) an AFM image and a surface profile of the GaAs/InGaAs/GaAs heterstructure surface modified by four beam standing wave-pulsed laser radiation with the pulse energy density of ≈ 1.5 J·cm^{-2} are presented. Note that in this experiment the bisector between the incidence angles of one pair of the beams was tuned off the normal to the sample surface. This resulted in the modulation of the intensity in the standing wave pattern maxima in the diagonal direction from the upper left corner of the AFM image to the lower right one; the modulation period was doubled relative to the basic period of the standing wave pattern. That is why the corresponding modulation of the surface modification pattern can be observed in the AFM image in figure 30 (a).

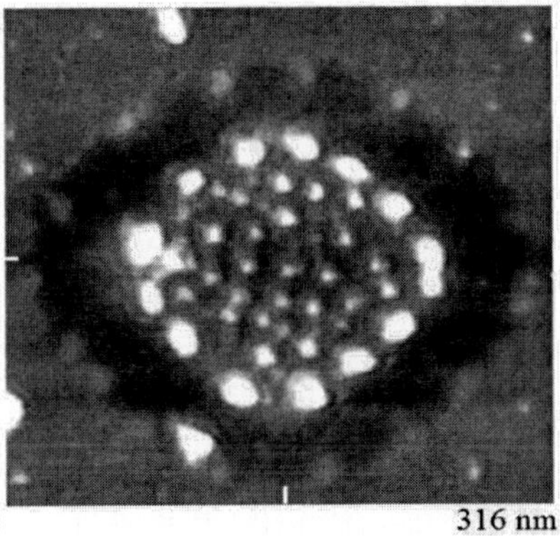

Figure 31. A typical AFM images of a modified area on the surface of GaAs/InGaAs/GaAs heterstructure.

In figure 31, a typical AFM image of a modified area on the GaAs/InGaAs/GaAs heterostructure surface is presented. Let us consider the following two features of the morphology of the modified heterostructure surface. First, the outer boundary of the modified areas (corresponding to the interference maxima) had a nearly rectangular shape, while the calculated intensity distribution near the standing wave maxima had a circular symmetry.

The second important feature is that a considerable number of islands appeared inside the modified areas, whereas no islands had been observed between these ones. It should be mentioned here that the general pattern of the modification on the surface of the strained heterostructure differs drastically from the one of the single crystal GaAs and Si as well as from the one of the amorphous films. In the case of the crystalline semiconductors, the single islands appeared near each interference maximum and a pit in the substrate was formed straight in the maxima (see Sec. A and B above). This was considered as the evidence of the crystalline material transfer. In the case of the amorphous metal films, the pits on the film surface surrounded by the swells (like in the case of an explosion) were formed (see Chapter IV, Sec. A).

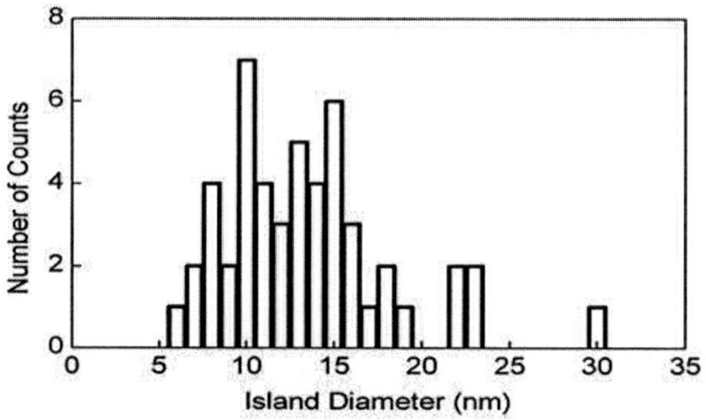

(a)

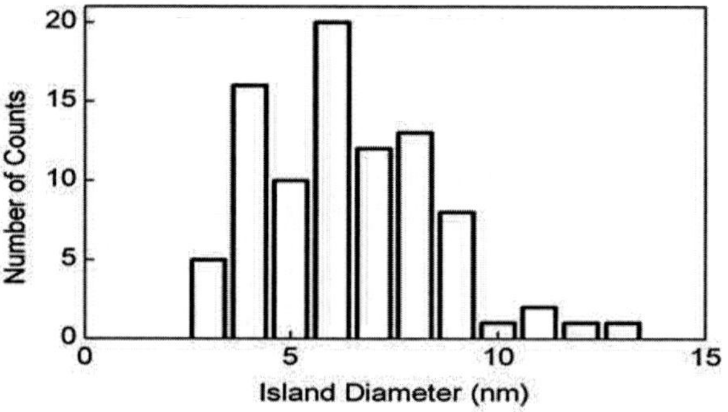

(b)

Figure 32. The histogram of the lateral size distribution for the peripheral (large) islands (a) and of the internal (small) ones (b) on the surface of GaAs/InGaAs/GaAs QW heterostructure modified by four-beam standing wave-pulsed XeCl laser radiation in the TE—TM polarization configuration.

The statistical analysis of the island sizes in the AFM images shown in figure 30 revealed a bimodal island size distribution. The sizes of the islands located in the periphery of the modified area (7 to 20 nm in diameter) differed significantly from the sizes of the islands located inside this area (3 to 8 nm, respectively). The size distributions of the peripheral (large) islands and of the internal (small) ones are presented in figure 32 (a) and (b), respectively. The surface densities of the peripheral islands and of the internal ones are close to 1×10^{10} cm^{-2} and to 5×10^{10} cm^{-2}, respectively.

From the experimental data presented here, one can make the following conclusions regarding the island formation process. First of all, under the conditions of these experiments, the ablation (i.e., removing the material from the processed surface) was absent. So far, the resulting surface morphology could be attributed to the material transfer on the surface only. The lateral sizes of the islands were 20 to 60 times smaller than the standing wave period(s). These sizes of the islands cannot be determined by a pure thermal effect since similar islands were obtained when the laser pulse energy density was increased two times.

As in the case of the laser modification of the single crystal GaAs and Si, understanding the modification process of the strained heterostructures requires consideration of the surface stresses as the driving forces of the nanoislands' self assembling. The process of island formation during the four-beam standing wave laser modification can be understood in terms of the Stranski-Krastanow growth. It is well known that the surface of a single crystal is in a stressed state. These surface stresses are related to the difference in the coordination of the atoms located in the bulk and on the surface of the crystal. The higher the quality of the single crystal surface, the greater the magnitude of surface stresses. Actually, the stresses relax at the defects of the single crystal surface that arise during growth and processing of a single crystal substrate. Any defects of any type known in crystallography can act as the stress relaxation centers. In this case, the spatial scale of the stress variations across the single crystal substrate surface depends on the quality of the substrarte. In the case of the substrate of high enough quality, the stress relaxation scale can be commensurable with the substrate dimensions. This is exactly what has been observed for the Si and GaAs single crystal substrates.

Unlike the homogeneous single crystals, the primary origin of the surface stresses in the heteroepitaxial films is the lattice mismatch between the film and the substrate. For InAs, the lattice constant $a_1 \approx 0.60584$ nm and for GaAs $a_2 \approx 0.56534$ (at 293K). The spatial scale of the stress variations is determined by the beats between the two lattice constants a_1 and a_2. In this approximation, one can easily obtain an expression for the minimum period of the spatial strain fluctuations to be $2a_1a_2/(a_1 - a_2)$. This estimate yields a value of ≈ 17 nm. In the samples used in the experiments described here, the averaged lattice constant of the $In_{0.35}Ga_{0.65}As$ alloy was ≈ 0.57952 nm (at 293K) that yields the period of the spatial variation of the elastic strain to be ≈ 51 nm. This value agrees fairly well with the averaged distance between the islands (presented in figure 30) measured experimentally by AFM. Such a coincidence can be treated as an additional argument that surface stresses play a key role in the standing wave laser assisted self assembling observed in the experiments described here. In this scope, the scatter in the distances between the adjacent islands can be attributed to the local variations in the InGaAs alloy composition.

Chapter 6

CONCLUSION

The results of the studies presented in this book demonstrate the capabilities and versatility of Four Beam Standing Wave XeCl-Pulsed Excimer Laser Lithography in the periodic patterning of various materials. These results supported by the theoretical calculations have demonstrated that using the four-beam interference of the high-power coherent pulsed-UV laser radiation one can form the nearly perfect periodic arrays of the nearly identical surface nanoobjects in an extremely wide range of the material systems. The lateral dimensions of these nanoobjects as well as the array periods in both directions can be controlled precisely by adjustment of the optical scheme of the interference (namely, of the angles of incidence of the laser beams onto the illuminated area) and of the beam parameters (such as the beam polarizations, the pulse energy, and the pulse duration). For the XeCl laser with the emission wavelength $\lambda = 308$ nm the minimum array period is ≈ 150 nm and as low relative uncertainty in the array period as $\sim 10^{-4}$ is achievable.

Although E-beam Lithography, X-Ray Lithography, and even Extreme UV Opitcal Lithography (again, the two latter techniques use the photomasks recorded by E-Beam Lithography) provide potentially much lower limits of the critical dimensions, the typical element size range of the UV Standing Wave Laser Lithography of 100 to 1000 nm covers a wide range of applications, first of all, in the integrated optics of the visible and of the near IR wavelength bands. Besides, using the nonlinear effects and/or the immersion can reduce the critical dimension achievable by Standing Wave Lithography drastically (down to $\sim$ 10 nm). Next, E-Beam Lithography is featured by infinite flexibility (i.e., any desired picture can be drawn by the electron beam scanning). This feature is inherited by X-Ray and UV

Lithographies along with the photomasks. In this scope, the capabilities of the Standing Wave Lithography are limited to the rather simple periodic patterns. However, in this particular case, Standing Wave Lithography takes an important advantage of the array period reproducibility. Last (but not least), Standing Wave lithography is much more cost effective compared to the other techniques listed above, especially when the direct surface modification is feasible. The studies presented in this book have demonstrated that even such hard material as diamond can be patterned using direct Standing Wave-Pulsed Laser Ablation. This method can be applied not only in the semiconductor diamond technology but also for secret marking of the diamonds used in jewelry for security and customs purposes, etc.

The process of modification of the solid surfaces in the direct Standing Wave-Pulsed Laser Modification is governed by the thermal processes near the surface. The theoretical investigations of the heat transfer in the thin films on the heat conducting substrates heated by a spatially periodic pulsed heat source helped us to find the conditions for the forming of the arrays of the surface nanostructures with necessary morphological parameters in various film and substrate material systems.

Application of the four-beam standing wave laser radiation to stimulate the self assembling of the semiconductor nanoislands opens new prospects in control of the self-assembling process and therefore in increasing the size uniformity of the nanoislands and the degree of their ordering in the substrate plane. The results of our experiments demonstrated the possibility of controlled self-assembling of the two-dimensional arrays of the nanoislands on the surface of single crystal Si(001) and GaAs(001) stimulated by pulsed standing wave UV laser radiation. Self assembling mechanism of the nanoisland formation provides the small nanoisland sizes down to $\approx$ 40 nm that is $\sim$ 10 times smaller than the standing wave period while the typical size of the nano relief features produced by direct Standing Wave Laser Ablation is $\approx$ 1/4 of the array period. So far, utilization of self-assembling reduces the lower limit for the size of the nanoislands formed by Standing Wave Laser Modification drastically. On the other hand, the standing wave pattern dictates the spatial arrangement of the nanoislands on the substrate and provides the uniform islands to arrange in nearly perfect two-dimensional superlattice. One can control both periods of this superlattice by adjusting the angles of incidence of the laser beams onto the patterned substrate. Also, the resulting nanoisland array morphology has been demonstrated to depend on the polarization of the interfering laser beams that opens additional opportunities for controlling the self-assembling process. The experiments on the four-beam standing wave-pulsed laser stimulation of the self-assembling on the surface of GaAs/InGaAs/GaAs(001) strained quantum well heterostructures and demonstrated that the interaction of the intrinsic strain field with the standing wave-induced thermal one results in a rather complex pattern of the surface nanostructures.

It should be stressed here that the four-beam laser setup could be combined quite easily with any growth chamber for MBE as well as for Vapor Phase Epitaxy just passing the laser beams through the appropriate viewports. Thus, standing wave-assisted self-assembling can be performed *in situ*, and the resulting nanoisland arrays could be built in any multilayered structure including the device ones. Taking into account all the above, we can conclude that the four-beam standing wave laser-assisted self-assembling is a very promising technique for various applications in nanotechnology.

REFERENCES

[1] Zaidi, S. H., Brueck, S. R. J., Hill, T. & Shagam, R. N. (1998). *Proc. SPIE, 3331*, 371-382.

[2] Brueck, S. R. J. (2005). *Proc. IEEE, 93*, 1704-1721.

[3] Zhang, L., Sugden, K., Williams, J. A. R., Bennion, I.., Reid, D. C. J. & Ragdale, C. M. (1995). *Opt. Lett. A, 20*, 1927-1929.

[4] Askins, C. C., Tsai, T. E., Williams, G. M., Putnam, M. A., Bashkansky, M. & Friebele, E. J. (1992). *Opt. Lett., 17*, 833-835.

[5] Luryi, S., Xu, J. & Zaslavsky, A. *Future Trends in Microelectronics: Up the Nano Creek*. Wiley-IEEE Press: Hoboken, NJ, 2007, 14.

[6] Bredikhin, V. I., Verevkin, Yu. K., Daume, E. Ya., Kuznetsov, S. P., Mal'shakova, O. A., Petryakov, V. N., Vostokov, N. V. & Polushkin, N. I. (2000). *Quantum Electron., 30*, 333-339.

[7] Savas, T. A., Farhoud, M., Smith, H. I. & Ross, C. A. (1999). *J. Appl. Phys., 85*, 6160-6165.

[8] Fernandez, A., Bedrossian, P. J., Baker, S. L., Vernon, S. P. & Kania, D. R. (1996). *IEEE Trans. Magn., 32*, 4472-4475.

[9] Smith, H. I. (2001). *Phys, E, 11*, 104-109.

[10] Zheng, M., Yu, M., Liu, Y., Liou, S. H., Sellmyer, D. J., Petryakov, V. N., Verevkin, Yu. K., Polushkin, N. I. & Salashchenko, N. N. (2001). *Appl. Phys. Lett., 79*, 2606-2608.

[11] Phillips, H. M., Callaham, D. L. & Sauerbrey, R. (1992). *Appl. Phys. A, 54*, 158-163.

[12] Ezaki, M., Kumagai, H. & Toyoda, K. (1993). *Jpn. J. Appl.Phys., 32*, 1308-1315.

[13] Pelissier, S., Biauc, D., Andrews, M. P., Najafi, S. I., Tishchenko, A. V. & Parriaux, O. (1999). *Appl. Opt., 38*, 6744-6746.

[14] Fernandez, A., Nguyen, H. T., Britten, J. A., Boyd, R. D., Perry, M. D., Kania, D. R. & Hawryluk, A. M. (1997). *J. Vac. Sci. Technol. B*, *15*, 729-734.

[15] Verevkin, Yu. K., Bronnikova, N. G., Korolikhin, V. V., Gushchina, Yu. Yu., Petryakov, V. N., Filatov, D. O., Bityurin, N. M., Kruglov, A. V. & Levichev. V. V. (2003). *Tech. Phys.*, *48*, 757-761.

[16] Verevkin, Yu. K., Klimov, A. Yu., Gribkov, B. A., Petryakov, V. N., Koposova, E. V. & Olaizola, S. M. (2008). *Quantum Electron.*, *38,* 1078-1082.

[17] Brundrett, D. L., Gaylord, T. K. & Glytsis, E. N. (1998). *Appl. Opt.*, *37*, 2534-2538.

[18] Xu, Y., Sun, H., Ye, J., Matssuo, S. & Misawa, H. (2001). *J. Opt. Soc. Am. B*, *18*, 1084-1090.

[19] *Nano-optoelectronics: Concepts, Physics, and Devices.* Grundmann, M., Ed., Springer: Berlin - Heidelberg, 2004.

[20] Ledentsov, N. N., Ustinov, V. M., Shchukin, V. A., Kop'ev, P. S., Alferov, Zh. I. & Bimberg, D. (1998). *Semicond.*, *32*, 343-361.

[21] Pchelyakov, O. P., Bolkhovityanov, Yu. B., Dvurechenskii, A. V., Sokolov, L. V., Nikiforov, A. I., Yakimov, A. I. & Voigtländer, B. (2000). *Semicond.*, *34*, 1229-1247.

[22] Teichert, C. (2002). *Phys. Rep.*, *365*, 335-369.

[23] Cerofolini, G. F., Arena, G., Camalleri, M., Galati, C., Reina, S., Renna, L., Mascolo, D. & Nosik, V. (2002). *Microel. Eng.*, *81***,** 405-429.

[24] *Semiconductor Nanostructures for Optoelectronic Applications.* Steiner, T. D., Ed., Artech House Semiconductor Materials and Devices Library, Artech House: Norwood, MA, 2004.

[25] Ustinov, V. M. (2004). *Semicond.*, *38*, 923-927.

[26] Kryzhanovskaya, N. V., Gladyshev, A. G., Blokhin, S. A., Musikhin, Yu. G., Zhukov, A. E., Maksimov, M. V., Zakharov, N. D., Tsatsul'nikov, A. F., Ledentsov, N. N., Werner, P., Guffart, F. & Bimberg, D. (2004). *Semicond.*, *38*, 833-835.

[27] Sgartato, A., Szkutnik, P. D., Balzarotti, A.; Motta, N. & Rosei, F. (2004). *Appl.Phys. Lett.*, *83*, 4002-4004.

[28] Shchukin, V. A., Borovkov, A. I., Ledentsov, N. N. & Kop'ev, P. S. (1995). *Phys. Rev. B*, *51*, 17767-17770.

[29] Capellini, G., De Seta, M., Spinella, C. & Evangelisti, F. (2003). *Appl. Phys. Lett.*, *82*, 1772-1774.

[30] Patella, F., Szkutnik, A. & Sgarlata, E. (2004). *J. Phys. Condens. Matter*, *16*, S1503-1505.

[31] Kitajima, T., Lu, B. & Leone, S. R. (2002). *Appl. Phys. Lett.*, *80*, 497-499.

[32] Zhoug, Z., Halilovic, A., Fromherz, T., Schäffler, F. & Bauer, G. (2003). *Appl. Phys. Lett.*, *82*, 4779-4781.

[33] Li, Q., Han, S. M., Bruek, S. R. J., Hersee, S., Jiang, Y. B. & Xu, H. (2003). *Appl. Phys. Lett.*, *83*, 5032-5034.

[34] Liu, F. & Lagally, M. G. (1997). *Surf. Sci.*, *386*, 169-175.

[35] Drozdov, Yu, N., Krasil'nik, Z. F., Lobanov, D. N., Novikov, A. V., Shaleev, M. V. & Yablonskiĭ, A. N. (2006). *Semicomd.*, *40*, 338-342.

[36] Pedraza, A. J., Fowlkes, J. D. & Guan, Y. E. (2003). *Appl. Phys. A*, *77*, 277-279.

[37] Verevkin, Yu. K., Daume, E. Ya., Petryakov, V. N., Gushchina, Yu. Yu. & Tikhov, S. V. (2005). *Tech. Phys. Lett.*, *31*, 759-762.

[38] Bauerle, D. (1996). *Laser Processing and Chemistry*. Springer: Berlin - Heidelberg.

[39] Mullendorn, M., Birkelund, K., Grey, F. & Madsen, S. (1996). *Appl. Phys. Lett.*, *69*, 3013-3015.

[40] Meltz, G., Morey, W. W. & Glenn, W. H. (1989). *Opt. Lett.*, *14*, 823-825.

[41] Askins, G. G., Tsai, T.-E., Williams, G. M., Putnam, M. A., Bashkausky, M. & Friebele, E. Y. (1992). *Opt. Lett.*, *17*, 833-835.

[42] Dong, L., Archambault, Y. L., Reekie, L., Russell, P. S. Y. & Payne, D. N. (1993). *Electron. Lett.*, *29*, 1577-1579.

[43] Fernandez, A. & Phillion, D. W. (1998). *Appl. Opt.*, *37*, 473-478.

[44] Baker, K. M. (1999). *Appl. Opt.*, *38*, 352-355.

[45] Meier, M., Mekis, A., Dodabalapur, A., Timko, A., Slusher, R. E., Yoannopoulos, Y. D. & Nalamasu, O. (1999). *Appl. Phys. Lett.*, *74*, 7-9.

[46] Ezaki, M., Kumagai, H., Toyoda, K. & Obara, M. (1993). *Jpn. J. Appl. Phys.*, *32*, 1308-1312.

[47] Seminogov, V. N., Panchenko, V. Ya. & Khudobenko, A. N. (1997). *JETP*, *111*, 174-181.

[48] Bychkov, Yu. I., Losev, V. F. & Panchenko, Yu. N. (1992). *Sov. J. Quantum Electron.*, *22*, 638-642.

[49] Neumann, A., Kuznetsova, Y. & Brueck, S. R. J. (2008). *Opt. Express*, *16*, 20477-20483.

[50] Perrone, M. R. & Yao, Y. B. (1994). *IEEE J. Quantum Electron.*, *30*, 1327-1330.

[51] Zhang, L. S., Williams, Y. A. R., Denion, Y., Reid, D. C. Y. & Ragdale, C. M. (1995). *Opt. Lett.*, *20*, 1927-1929.

[52] Othonos, A. & Lee, X. (1995). *Rev. Sci. Instr.*, *66*, 3112-3116.
[53] Veiko, S. M. & Metev, S. M. (1991). *Laser Technologies in Microelectronics*. Sofia: Bulgarian Academy of Science Publishing.
[54] Phillips, H. M., Callahan, D. L., Sauerbrey, R., Szabo, G. & Bor, Z. (1992). *Appl. Phys. A*, *54*, 158-161.
[55] Chung, Y. & Shi-Woo, R. (2001). *J. Mater. Sci. Mater. Electron.*, *12*, 697-700.
[56] Lee, M., Moou, S., Haatana, M. & Grigoropoulos, C. P. (2001). *Appl. Phys. A*, *73*, 317-321.
[57] Efremov, M. D., Bolotov, V. V., Volodin, V. A., Kochubei, S. A. & Kretinin, A. V. (2002). *Semicond.*, *36*, 102-106.
[58] Alekseev, A. M., Verevkin, Yu. K., Vostokov, N. V., Petryakov, V. N., Polushkin, N. I., Popkov, A. F. & Salashchenko, N. N. (2001). *JETP Lett.*, *73*, 192-196.
[59] Boyd, R. D., Britten, Y. A., Decker, D. E., Shore, B. W., Stuart, B. C., Perry, M. D. & Li, L. (1995). *Appl. Opt.*, *34*, 1697-1700.
[60] Verevkin, Yu. K. & Daume, E. Ya. (1998). *Opt. Spectrosc.*, *85*, 239-244.
[61] Wang, S., Zhou, C., Zhang, Y. & Ru, H. (2006). *Appl. Opt.*, *45*, 2567-2569.
[62] *Silicon Photonics*. L. Pavesi, & D. J. Lockwood, (Eds.), Springer: Berlin - Heidelberg, 2004.
[63] Vlasov, S. N. & Koposova, E. V. (2003). *Radio Phys. Quantum Electron.*, *46*, 5-14.
[64] Dragomir, N. M., Rollinson, C. & Wade, S. A. (2003). *Opt. Lett.*, *28*, 789-791.
[65] Park, D. & Kim, M. (2004). *Opt. Lett.*, *29*, 1849-1851.
[66] Haro-Poniatowski, E., Fort, E., Lacharme, J. P. & Ricolleau, C. (2005). *Appl. Phys. Lett.*, *87*, 3103-3105.
[67] Guo, L. J. (2004). *J. Phys. D Appl. Phys.*, *37*, R123-159.
[68] Pisignano, D., Maruccio, G. & Mele, E. (2005). *Appl. Phys. Lett.*, *87*, 3109-3111.
[69] Heon, L., Hong, S. & Yang, K. (2006). *Appl. Phys. Lett.*, *88*, 3112-3114.
[70] Kanamori, Y., Sasaki, M. & Hane, K. (1999). *Opt. Lett.*, *24*, 1422-1424.
[71] Kintaka, K., Nishii, J., Mazutani, A., Kikuta, H. & Nakano, H. (2001). *Opt. Lett.*, *26*, 1642-1644.
[72] Rastei, M. V., Meckenstock, R., Bucher, J. P., Devaux, E. & Ebbessen, T. (2004). *Appl. Phys. Lett.*, *85*, 2050-2052.
[73] Kim, D. (2005). *Appl. Opt.*, *44*, 3213-3217.

[74] Zeng, L., Hong, Y., Liu, J., Feng, N., Duan, X., Kimerling, L. & Alamariu, B. (2006). *Appl. Phys. Lett.*, *89*, 1111-1113.

[75] Guise, P., Yates, J., Levy, J., Ahner, J., Vaithynathan, V. & Schlom, D. (2005). *Appl. Phys. Lett.*, *87*, 1902-1904.

[76] Zhong, Z., Katsaros, G., Stoffel, M., Costantini, G., Kern, K., Schidt, G., Jin-Phillipp, N. & Bauer, G. (2005). *Appl. Phys. Lett.*, *87*, 3102-3104.

[77] Zaidi, S. H. & Brueck, S. R. J. (1999). *Proc. SPIE, 3676*, 371–378.

[78] Brueck, S. R. J. In International Trends in Applied Optics. International Commission on Optics. SPIE Press: Bellingham, WA, 2002, 85–110.

[79] Hoffnagle, J. A., Hinsberg, W. D., Houle, F. A. & Sanchez, M. I. (2003) *J. Photopolym. Sci. Technol., 16*, 373–379.

[80] Naqvi, S. S. H., Zaidi, S. H., Brueck, S. R. J. & Mc.Neil, J. R. (1994). *J. Vac. Sci. Technol. B, 12*, 3600–3606.

[81] Burnett, J. H. & Kaplan, S. G. (2004). *J. Microlith. Microfab. Microsys., 3*, 68–72.

[82] International Technology Roadmap for Semiconductors [Online]. http://public.itrs.net.

[83] Biswas, A. & Brueck, S. R. J. (2004). *J. Microlith. Microfab. Microsys., 3*, 35–43.

[84] Polushkin, N. I., Gusev, S. A., Drozdov, S. A., Verevkin, Yu. K. & Petryakov, V. N. (1997). *J. Appl. Phys.*, *81*, 1-3.

[85] Verevkin, Yu. K., Petryakov, V. N. & Polushkin, N. I. (1998). *Tech. Phys. Lett.*, *24*, 460-463.

[86] Moser, A,. Takano, K., Margulies, D. T., Albrecht, M., Sonobe, Y., Ikeda, Y., Sun, S. & Fullerton, E. E. (2002). *J. Phys. D: Appl. Phys., 35*, R157–R167.

[87] Chou, S. Y., Wei, M. S., Krauss, P. R. & Fischer, P. B. (1994). *J. Appl. Phys., 76,* 6673–6675.

[88] Chou, S. Y. (1997). *Proc. IEEE, 85,* 52–671.

[89] Meier, G., Kleiber, M., Grundler, D., Heitmann, D., & Wiesendanger, R. (1998). *Appl. Phys. Lett., 72*, 2168–2170.

[90] Haast, M. A. M., Schuurbuis, J. R., Abelmann, L., Lodder, J. C. & Popma, T. L. (1998). *IEEE Trans. Magn., 34*, 1006–1008.

[91] Theilen, M., Kirsch, S., Weinforth, H., Carl, A. & Wassermann, E. F. (1998). *IEEE Trans. Magn., 34*, 1009–1011.

[92] Ross, C. A., Haratani, S., Castaño, F. J., Hao, Y., Hwang, M., Shima, M., Cheng, J. Y., Vögel, B., Farhoud, M., Walsh, M. & Smith, H. I. (2002). *J. Appl. Phys., 91*, 6848–6853.

[93] Parkin, S. S. P., Roche, K. P., Samant, M. G., Rice, P. M., Beyers, R. B., Scheuerlin, R. E., O'Sullivn, E. J., Brown, S. L., Bucchigano, J., Abraham, D. W., Lu, Y., Rooks, M., Trouilloud, P. L., Wanner, R. A. & Gallagher, W. J. (1999). *J. Appl. Phys., 85*, 5828–5833.

[94] Everitt, B. A., Pohm, A. V., Beech, R. S., Fink, A. & Daughton, J. M. (1999). *IEEE Trans. Magn., 34,* 1060–1062.

[95] Zhu, M., Grütter, P., Hao, Y., Castaño, F. J., Haranti, S., Ross, C. A., Vögel, B., & Smith, H. I. (2003). *J. Appl. Phys., 93*, 1132–1136.

[96] Metlushko, V., Welp, U., Crabtree, G. W., Osgood, R., Bader, S. D., De Long, L. E., Zhang, Z., Brueck, S. R. J., Ilic, B., Chung, K. & Hesketh, P. J. (1999). *Phys. Rev. B, 60,* 12585–12592.

[97] Field, S. B., James, S. S., Barentine, J., Metlushko, V., Crabtree, G., Shtrikman, H., Ilic, B. & Brueck, S. R. J. (2002). *Phys. Rev. Lett., 88*, 067003.

[98] Veselago, V. G. (1968). *Sov. Phys. Usp., 10*, 509–514.

[99] Pendry, J. B. (2000). *Phys. Rev. Lett., 85*, 3966–3969.

[100] Zhang, S., Fan, W., Malloy, K. J., Brueck, S. R. J., Panoiu, N. C. & Osgood, R. M. (2005). *Opt. Express, 13*, 4922–4931.

[101] Ku, Z., Brueck, S. R. J. (2007). *Opt. Express, 15*, 4515-4522.

[102] Yen, T. J., Padilla, W. J., Fang, N., Vier, D. C., Smith, D. R., Pendry, J. B., Basov, D. N. & Zhang, X. (2004). *Science, 303*, 1494–1496.

[103] Fan, X., Demaree, D. P., St.John, J. M., Tripathi, A. & Oliver, S. R. J. (2008). *Appl. Phys. Lett., 92*, 193108.

[104] Menzies, D. B., Dai, Q., Bourgeois, L., Caruso, R. A., Cheng, Y.-B., Simon, G. P. & Spiccia, L. (2007). *Nanotechnology, 18,* 125608.

[105] Al-Dmour, H., Taylor, D. M. & Cambridge, J. A. (2007). *J. Phys. D: Appl. Phys,. 40,* 5034–5038.

[106] Lin, C.-J., Yu, W.-Y. & Chien, S.-H. (2007). *Appl. Phys. Lett., 91*, 233120.

[107] Lin, C.-J., Yu, W.-Y. & Chien, S.-H. (2008). *Appl. Phys. Lett., 93*, 133107.

[108] Tien, T.-C., Pan, F.-M., Wang, L.-P., Lee, C.-H., Tung, Y.-L., Tsai, S.-Y., Lin, C., Tsai, F.-Y. & Chen, S.-J. (2009). *Nanotechnology, 20*, 305201.

[109] Wu, Y., Jia, W., An, Q., Liu, Y., Chen, J. & Li, G. (2009). *Nanotechnology, 20*, 245101.

[110] De Nyago, T., Wang, J., Wu, N. & Lewis, J. P. (2009). *Appl. Phys. Lett., 94*, 093101.

[111] Eufinger, K., Poelman, D., Poelman, H., de Gryse, R., & Marin, G. B. (2007). *J. Phys. D: Appl. Phys.*, *40*, 5232-5237.

[112] Yu, H., Yu, J., Cheng, B., & Liu, S. (2007). *Nanotechnology*, *18*, 065604.

[113] Chuang, H.-Y. & Chen, D.-H. (2009). *Nanotechnology*, *20*, 105704.

[114] Lin, C.-J., Lu, Y.-T., Hsieh, C.-H. & Chien, S.-H. (2009). *Appl. Phys. Lett.*, *94*, 113102.

[115] Chi, C.-F., Liau, S.-Y. & Lee, Y.-L. (2010). *Nanotechnology*, *21*, 025202.

[116] Dai, K., Peng, T., Ke, D. & Wei, B. (2009). *Nanotechnology*, *20*, 125603.

[117] Das, P. P., Mohapatra, S. K. & Misra, M. (2008). *J. Phys. D: Appl. Phys.*, *41*, 245103.

[118] Bredikhin, V. I., Burenina, V. N., Verevkin, Yu. K., Kirsanov, A. V., Petryakov, V. N., Vostokov, N. V., Dryakhlushin, V. F. & Klimov, A. Yu. (2004). *Tech. Phys.*, *49*, 1191-1197.

INDEX

A

B

C

G

H

I

O

P

Q

R

S

T

U

V

W